Live, Learn and Be Happy With Epilepsy

By Stacey Chillemi

LULU EDITION

PUBLISHED BY:

Stacey Chillemi on Lulu

Live, Learn and Be Happy With Epilepsy

Table of Contents

Live, Learn and Be Happy With Epilepsy

Learning about Your Disorder and How to Cope with It

DEDICATION

This book is dedicated to all the people who suffer from epilepsy. I struggled many years trying to learn how to live a happy, fulfilling and productive life living with epilepsy. I hope my book helps you and gives you new insight on how to cope with epilepsy. My support and love go out to all of you.

ACKNOWLEDGEMENTS

Writing this book was a project that I could not have completed by myself. I created this book using a combination of thoughts; ideas, experiences, and lives that have help change my life. My motivation to write this book began when I walked into a bookstore and was appalled to find so few books on the topic of epilepsy. It made me even angrier to find that doctors seemly wrote a majority of the books for doctors. The vocabulary and the approach made them difficult for the reader to understand.

At that point, I knew that I wanted to change the way society looked at people with epilepsy, and how people with epilepsy looked at them. I wanted people to understand what epilepsy is and what it does to people who suffer with the disorder. Therefore, I decided to write a book about epilepsy recalling my own and other people's personal experiences with the disorder. I wanted to help others with this book and educate individuals who knew nothing about epilepsy.

I cannot even begin to express my thanks to all the people who helped me with this book. I first must give appreciation to my husband Michael and my son Mikey and my daughter Alexis, who gave me encouragement to write this book. You stood by my side through all the difficult times I had dealing with epilepsy. To all my family, friends and loved ones who, *"Live, Learn and Be Happy with Epilepsy,"* has played a big role in my life and has supported all my decisions that I made concerning epilepsy. I would like to express my deepest gratitude to all of you.

FOR THE MILLIONS OF PEOPLE WHO WANT TO LEARN HOW

TO LIVE
A HEALTHY, HAPPY
AND PRODUCTIVE LIFE
WITH EPILEPSY

- **Stacey Chillemi**

FORWARD

Why do so many people feel life is over just because they developed epilepsy? Why do so many people with epilepsy carry so much anger inside themselves? Emotions similar to the ones I just mentioned are common with people who have epilepsy. At one time or another, everyone with epilepsy has experienced some type of anger or frustration. We all have a purpose in life. We have a plan, a future that lies ahead.

I am here to help you, explain to you, and guide you, so you understand and learn how to live a happy and productive life with epilepsy. The techniques in "Live, Learn, and Be Happy with Epilepsy," will help the reader build the inner power to do anything or become anything they want in life. The approaches for dealing with epilepsy will enable the reader to reform a better direction in their everyday life of living and dealing with their disorder. Most important, this book shows the reader that you can live a happy and productive life with epilepsy.

The tools in this book will help build confidence. Once the reader establishes self-assurance they reader will start to see their inner strength boost. When one quality improves, all their other attributes will enhance also. This book will give the reader the tools to learn how to cope with their epilepsy disorder, so the reader can live a positive productive life. The reader can make their life anything they want if they have positive goals to focus on and if they have a good understanding of how to approach them.

This book gives the reader the tools they need to gain encouragement and strength to overcome having epilepsy and being able to live life to its fullest. When I wrote this book, I wanted to be able to focus on certain topics related to epilepsy that not many authors and doctors have discussed. One of the main goals in this book is to help the reader recognize that life has much to offer. Life does not have to stop just because you have epilepsy. In this book, we will be discussing what people go through emotionally when they have epilepsy. We will be going over different ways to help the reader emotionally, physically and spiritually.

The aim in this book is to show readers how to live with epilepsy, empowering them to take responsibility for their life and well-being. Unlike other books on epilepsy, **"Live,**

Learn and Be Happy," does not focus on the same subject matter. Instead, it tries to motivate people with epilepsy, urging readers to look at life and live life in a positive and productive manner. While seemingly revolutionary, the message is simple: It is important that people with epilepsy learn how to live with epilepsy and endure it. Everything you do and say affects the people around you.

Eventually if people with epilepsy do not learn how to deal with all these issues, they could end up really destroying themselves emotionally, physically and spiritually. One way to prevent this from happening is to develop a lifestyle that is suitable for your own needs. The reader needs to make sure that they create a lifestyle that is going to make them happy over time. They need to be their own designer, creating pathways to a fulfilling future. There is a whole world in front of you. This world has millions of opportunities just waiting for you to encounter. It does not matter what age you are. You can achieve anything in life that you put your mind too. Do not think that just because you have epilepsy you are different from anyone else because you are not any different. In Fact, you are probably better!

INTRODUCTION

Epilepsy Affects Millions of People Worldwide, With More Than Two Million People in the United States Suffering from the Disorder

"Live, Learn, and Be Happy with Epilepsy," is a refreshing new look on how to cope with your epilepsy and journey back to happiness. Epilepsy needs to be understood from point of view of someone who has the disorder. We need to go inside ourselves and understand how we our feeling on the inside.

I have created an approachable, appealing, more compelling way to help people with epilepsy cope with their disorder. Unlike other books on epilepsy, it will focus on the connection between the mind, body, and spirit as it relates to epilepsy, showing readers how to use that connection to cope with their epilepsy. The tools in this book are simple to learn and the results are.

Individuals of all ages suffer from epilepsy and seizures. People with epilepsy are not the only ones who suffer, family and friends suffer. How would you feel if you walked into the house to find the one you loved or felt close to on the floor with their eyes rolled back and their body shaking uncontrollably? It is something you do not want to experience. These individuals are yearning to learn more about the disorder that is disrupting their life and they are unable to do it because most existing books on the market approach the subject from "The physician's point of view." There is little emphasis made on showing people with epilepsy that you can make changes in your life to live positively with epilepsy.

The following chapters will deal with the mind, body, and spirit connection. The book includes chapters on the role of motivation, self-esteem, and faith. In order to help readers reduce stress and learn how to incorporate epilepsy in their life so they can live with the disorder feeling good about them.

This book will show you how you can create your own *"personal journal"* that combines use of positive imagery, relaxation, meditation, and exercise. The book will include my personal experience growing up with epilepsy. The book will also talk about how I overcame the disorder and got on with my life. The end of the book will include a section discussing how to keep you in good health emotionally, physically and spiritually.

My goal is to help readers understand that medicine is not the only way to keep seizures under control. You need to incorporate a healthy lifestyle as well. You need to keep healthy by eating right, exercising, and sleeping properly.

This chapter will introduce readers to techniques and exercises that will help you put the advice of the chapter into immediate practice. The last chapter of the book will include what medical technology has in store for people with epilepsy, discussing new medicines, treatments, and surgical procedures. In addition, the book will include a glossary to help understand medical terms in the book that may be unfamiliar to the reader. Through example and encouragement will offer readers a variety of strategies for coping with epilepsy. It is a workable program for coping with epilepsy and forming a healthy relationship with their mind, body and spirit enabling readers to overcome their disorder and get on with their lives.

SECTION 1: THE HISTORY OF EPILEPSY

Hippocrates

(460 - 377)

Galen

(129 - 200)

A. of Tralleis

(525 - 600)

Avicenna

(980 - 1037)

The Middle Ages

The Renaissance

Paracelcus

(1439 - 1541)

S.A. Tissot

(1728 - 1797)

J.H. Jackson

(1935 - 1911)

When did epilepsy begin? When epilepsy recognized? What did people think about epilepsy? What famous people had epilepsy? In Section 1, I describe the history of epilepsy and the correct and incorrect statements that people thought about epilepsy.

EPILEPSY

The Most Important Secrets You Must Learn In Order To Live, Learn, and Be Happy With Epilepsy

Chapter 1: The History of Epilepsy

Epilepsy has been on this planet as far back as time will take us. Epilepsy has existed since the world was blessed with the birth of humans and it has been recognized since the earliest medical writings. Epilepsy has attracted much attention in the medical community, since the beginning of time. Few medical conditions have attracted so much attention and generated so much debate as epilepsy.

460 - 375 B.C.

Hippocrates was a physician from the island of Cos in ancient Greece. Known as the "Father of Medicine." The Greek physician Hippocrates wrote the first book on epilepsy, On the Sacred Disease.

"It is thus with regard to the disease called sacred: it appears to me to be in no way more divine nor more sacred than other diseases. The brain is the cause of this affliction. When the phlegm from the brain runs down through the veins, the patient loses his speech and foams at the mouth, his hands are contracted, the eyes contorted, he becomes insensible, and in some cases, the bowels are emptied. The patient kicks with his feet. The patient must endure all these symptoms when the cold phlegm flows into the warm blood."

Correct statements:

- ✓ Epilepsy is a natural disease, not a "sacred" one.
- ✓ Seizures begin in the brain.

Incorrect statement:

Epilepsy motion is caused by surplus phlegm - Hippocrates'

humoral theory. Disagreeing with the idea that epilepsy is a curse or a visionary power, Hippocrates proves the truth: It is a brain disorder. "It is with looked upon as the disease called Sacred". Hippocrates said, "It appears to me to be nowise more divine nor more sacred than other diseases, but has a natural cause like other affections.

"The Greek physician Hippocrates wrote the first book on epilepsy, titled *On the Sacred Disease*, around 400 BC. Hippocrates recognized that epilepsy was a brain disorder, and he spoke out against the ideas that seizures were a curse from the gods and that people with epilepsy held the power of prophecy.

Unfortunately, false ideas die slowly in our society, and for centuries epilepsy was viewed as a curse of the gods, or worse. For example, a 1494 handbook on witch-hunting, *Malleus Maleficarum*, written by two Dominican friars under papal authority, said that one of the ways of identifying a witch was by the presence of seizures. This book guided a wave of persecution and torture, which caused the deaths of more than 200,000 women thought to be witches.

Misinterpretation has continued for many years. In the early 19th century, people who had critical epilepsy and people with psychiatric disorders were cared for in asylums, but the two groups were kept separated because seizures were thought to be contagious. In the early 1900s, some U.S. states had laws forbidding people with epilepsy to marry or become parents, and some states permitted sterilization.

The modern medical age of epilepsy began in the mid-1800s, under the leadership of three English neurologists: Russell Reynolds, John Hughlings Jackson, and Sir William Richard Gowers. Hughlings Jackson is meaning of a seizure as "an occasional, an excessive, and a disorderly discharge of nerve tissue on muscles is still in existence today." Hughlings Jackson also taught others that seizures could alter consciousness, sensation, and behavior.

The past century has brought an abrupt increase of information and research about the functions of the brain and about epilepsy. Epilepsy research continues at an energetic speed, with research ranging from how microscopic particles and channels in the cell trigger seizures, to the Progress of new seizure medicines, and to a better understanding of how epilepsy affects social and intellectual development.

In 70 A.D. Gospel According to Mark (9:14-29), Jesus Christ casts out a devil from a young man with epilepsy: "Teacher, I brought you my son, who is possessed by a spirit that has robbed him of speech. Whenever it seizes him, it throws him to the ground. He foams at the mouth, gnashes his teeth, and becomes rigid. I asked your disciples to drive the spirit out, but they could not." (NIV)

Roman medicine: Galen (129 - approx. 200):

On diseased parts of the body

Born AD 129, Pergamum, Mysia, Anatolia

died *c.* 216

Greek physician, writer, and philosopher: Latin Galenus

He became chief physician to the gladiators in AD 157. Later, in Rome, he became a friend of Marcus Aurelius and physician to Commodus. Galen saw anatomy as essential, based on animal experiments, described cranial nerves and heart valves, and showed that arteries carry blood, not air. However, in extending his findings to human anatomy, he was often in error.

Following Hippocratic concepts, he believed in three connected body systems—brain and nerves for sensation and thought, heart and arteries for life energy, and liver and veins for nutrition and growth—and four body fluids—blood, yellow bile, black bile, and phlegm—that needed to be in balance. Few had the skills to challenge his seductive physiological theory. He wrote about 300 works, of which about 150 survive. As they were translated, his influence spread to the Byzantine Empire, Arabia, and then Western Europe. A stimulation of interest in the 16th century led to new anatomical investigations, which caused the overthrow of his ideas when Andreas Vesalius found anatomical errors and William Harvey correctly explained blood circulation.

According to Galen there are three forms of epilepsy: "In all forms it is the brain which is diseased; either the sickness originates in the brain itself, or it rises in sympathy into the brain from the cardiac orifice of the stomach. Seldom, however, it can have its origin in any part of the body and then rises to the head in a way, which the patient can feel.

Case description: I heard the boy say that his condition began in his lower leg and then moved up through the thigh, the groin and side of the chest above the affected thigh up to the neck and then to the head. As soon as [the condition] reached this part, he said that he was no longer aware of himself. When the doctors asked what the movement into the head was like, [another] boy said the movement upwards was like a cold breeze (aura)."

Correct statements:

- ✓ "The brain is diseased."
- ✓ There are signs of the onset of a seizure, which only the patient is aware of: the **aura**. This is the first time this term is used in medical literature.

Incorrect statement:

- ✓ Epileptic activity can (primarily) originate in one part of the body and then (secondarily) affect the brain. (*Correct*: Every seizure begins primarily in the brain!)

Byzantine medicine: ***Alexandros of Tralleis*** *(approx. 525 - 605):*

12 books about medicine

"The proof that epilepsy begins in the stomach lies in the fact that a feeling of restlessness and gnawing begins in the stomach and then the patient feels the affliction approaching. As soon as the patient gets up in the morning and has emptied his bowels, he should drink an infusion of hyssop, which will do him a lot of good, as many have been healed simply by drinking this, and were only taken ill two or three times. It is forbidden to drink undiluted wine after taking a bath as nothing can set off a seizure more easily than this - and indeed undiluted wine is in general dangerous for all epileptics."

Correct statements:

- ✓ An aura (**Galen**) can take the form of a stomach complaint.
- ✓ Alcohol can increase the risk of having an epileptic seizure.

Incorrect statement:

- ✓ Plants or parts of plants (like hyssop) are effective cures for epileptic seizures. (*Correct*: There are no herbal remedies for epilepsy.)

Arab medicine: Avicenna (980-1037): Canon medicinae Born 980, Bukhara, Iran died 1037, Hamadan Islamic philosopher and scientist.

Arabic ***Ibn Sina*** *in full* ***Abu 'Ali al-Husayn ibn 'Abd Allah ibn Sina*** He became physician to several sultans and also twice served as vizier. His *Canon of Medicine* was long a standard work in the field. He is known for his great encyclopedia of philosophy, *The Book of Healing*. His other writings include *The Book of Salvation* and *The Book of Directives* and *Remarks*. His interpretations of Aristotle influenced European Scholasticism. His system rests on a conception Of God as the necessary existent: only in God do essence, (what God is) and existence (that God is) coincides.

"Epilepsy is a disease which prevents those organs affected from using the senses, moving and walking upright and this is caused by a blockage. Usually it is a general seizure, caused by some damage, which affects the front cerebral ventricle; and it is impossible for the person affected to remain standing upright.

Correct statement:

- ✓ Epileptic seizures originate in the brain and often lead to loss of upright posture (falling) and "impairment of the senses" (e.g. twilight states or unconsciousness).

Incorrect statement:

- ✓ Seizures caused by epilepsy are caused "mechanically" by a blockage and are localized in the front cerebral ventricle. (*Correct*: Epileptic seizures are the result of a stimulation disturbance in the cerebral nerve cells.)

Medicine in the Middle Ages: Falling Sickness Blessing (14th/15th century):

As convulse and bewitch are walking across the heath, they meet the Holy Virgin Mary, the Virgin Mary asks convulse and bewitch: convulse and bewitch, where are you going?' Convulse and bewitch say: 'We are going to him and him.'

The Virgin Mary asks, 'What are you going to do there? Convulse and bewitch say, 'We're going to tear flesh, drink blood and break legs.'

The Virgin Mary says, 'you must not do that: you must go where there are bare rocks, there you can tear flesh, drink blood and break legs.' May god the father, god the son and god the Holy Ghost help us. Amen."

Correct idea:

Confidence, positively and courage are a good basis for the successful treatment of epilepsy.

Incorrect statement:

Epilepsy is a "bewitched" disease, which can only be cured with divine aid.

Correct:

Epilepsy is an organic disease that can be treated using rational means of therapy. In 1494, a handbook on witch-hunting, Malleus Maleficarum, brings a wave of persecution and torture, leading to the death of more than 200,000 women. Written by two Dominican friars under papal authority, the book identifies the presence of seizures as a characteristic of witches.

The Renaissance *(approx. from 1500):*

The Renaissance physician (around 1500) gradually rejects the medical beliefs of the Middle Ages, which were influenced by the Christian faith and by superstition; he begins to make tentative steps towards practicing natural "scientific" medicine. One of the most famous and influential Renaissance physicians was Paracelcus, who devoted much time to the study of the "falling sickness", epilepsy.

The painting shows the Renaissance physician fighting death and his companions-diseases (depicted as black birds) with "new weapons":

The Stag Chemist's, Offenburg: Mural painting around 1900 using Renaissance motifs.

Correct statement

- ✓ With scientific thought symbolized by the owl

With the experience and knowledge of ancient Greek and Roman physicians (symbolized by books);

Incorrect Statement

- ✓ Natural, specially prepared substances (symbolized by the mortar and pestle);

With chemical-physical experiments (symbolized by chemical apparatus).

Medicine in the Renaissance: Paracelcus (1493 - 1541):

On ailments which rob us of our reason (1525)

"'And such falling sicknesses have five seats: One is in the brain, the second in the liver, the third in the heart, the fourth in the intestines, the fifth in the limbs. [...] And this is not only so in human beings but also in every living creature, in animals, which also fall down in the same form as in humans, and the earthquake also has the same origin as the falling sickness. We say that it is impossible to cure the root of the disease, but that it is possible to prevent the root from growing. "

Correct statements:

- ✓ Epilepsy is organic. It is not an unnatural, mystical disease.
- ✓ Animals can also have epilepsy.

It is not always possible to cure the cause ("*root*") of the disease, but the symptoms can be treated ("*prevent the root from growing*"): the principle of symptomatic therapy.

Incorrect statements:

- ✓ Epilepsy can have its seat in the liver, the heart, in the intestines or in the limbs. (*Correct*: Every epileptic seizure originates in the brain.)

Earthquakes are also of an epileptic nature.
(*Correct*: Epileptic activity is connected with the nerve cells.)

Medicine in the 18th century: Samuel Auguste A. D. Tissot

(1728 - 1797):

Treatise on Epilepsy or the Falling Sickness (1771)

"In order to be in a position to cure this disease, one must first take pains to examine whether there is any sympathetic cause which supports it, and what this could be; or whether it is an idiopathic one, that is to say whether it simply stems from an over-sensitivity of the brain.

At last, valerian has fortunately become the favorite remedy of all sensible physicians. I am convinced that, if this does not have an effect, then it is because the malady is incurable."

Correct statement:

Differentiation between "idiopathic" and "sympathetic" epilepsies.

Idiopathic: Epilepsy is mainly caused by an inherent tendency to the disease.

Sympathetic (symptomatic): The epilepsy is a symptom of a

primary disease (e.g. brain tumor, metabolic disturbance, cerebral scarring after injury).

Incorrect statement:

Valerian is a good remedy for epilepsy. (*Correct*: Valerian can have a calming effect, but does not suppress seizures.)

In 1859-1906, under the leadership of three English neurologists--John Hughlings Jackson, Russell Reynolds, and Sir William Richard Gowers--the modern medical era of epilepsy begins. In a study, Jackson defines a seizure as "an occasional, an excessive, and a disorderly discharge of nerve tissue on muscles." He also recognizes that seizures can alter consciousness, sensation, and behavior. "The fit usually begins, it is to be observed, in that part of the face, of the arm, and oft the leg, which has the most varied uses. The fits, which begin in the hand, begin usually in the index finger and thumb; fits, which begin in the foot, begin usually in the great toe.

It may be that the order of frequency mentioned point merely to an order of frequency in liability of parts to become diseased. Parts which have the most varied uses will be represented in the central nervous system by most ganglion cells."

All of Jackson's statements are correct!

In 1904, the term "epileptologist" was first used to describe a person who specializes in epilepsy. William Spratling, the neurologist who coined the word, is regarded as North America's first epileptologist.

In 1912, two groups of chemists on their own created Phenobarbital under the name of Luminal. Phenobarbital is the oldest AED in common clinical use.

In 1920, the ketogenic diet is one of the oldest forms of treatment for epilepsy. Created in the 1920s when there were few effective treatments for epilepsy, this special diet, which is high in fat, low in protein, and has negligible amounts of carbohydrate, was created to simulate some of the metabolic effects of fasting, a state known to decrease seizures in some individuals.

In 1929, a German psychiatrist named Hans Berger announced to the world that it was possible to record electric currents generated on the brain, without opening the skull, and to illustrate them graphically onto a strip of paper. Berger named this new form of recording as the electroencephalogram (EEG).

In 1939, Discovery and clinical testing of phenytoin (PHT) by Merritt and Putnam introduced both a major new non-sedating AED and an animal model of epilepsy. For over forty years, PHT has been a first-line medication for the prevention of partial and tonic-clonic seizures and for the acute treatment of seizures and status epilepticus.

In 1953, Schindler at Geigy synthesized Carbamazepine (CBZ) in an effort to try to compete with the new introduced antipsychotic, chlorpromazine. Over the years, CBZ has gained acceptance as a first-line treatment for partial and tonic-clonic seizures.

In 1958, Ethosuximide (ESM) was introduced as an AED and has been the drug of choice for children with absence seizures who do not also have tonic-clonic or myoclonic seizures. ESM is also effective for atypical absence seizures.

In 1963, Sodium Valproate (VPA) anticonvulsant property was recognized serendipitously when it was used by Pierre Eymard as a solvent for a number of other compounds. VPA is effective over the complete range of seizures.

In 1968, The Epilepsy Foundation is a national, charitable organization, founded in 1968 as the Epilepsy Foundation of America. The only such organization devoted to the well-being of people with epilepsy, there objective is to work for children and adults affected by seizures through research, education, advocacy and service.

There national office is in Landover, MD, a suburb of Washington, D.C. More than 60 affiliated Epilepsy Foundations serve people with seizures, and their families, in hundreds of communities nationwide.

A volunteer board of directors governs their work; a distinguished board of physicians and scientists oversees the scientific and medical programs. The Foundation is a member of the National Health Council and the International Bureau for Epilepsy. The organization is supported almost entirely by private donations.

In 1970, the Veterans Administration spearheads a movement toward creating epilepsy centers, introducing a new breed of neurologists who began to specialize in the treatment and research of epilepsy.

In 1990, even in the twentieth century, some U.S. states had laws forbidding people with epilepsy to marry or become parents, and some states permitted sterilization. To establish a clear and comprehensive prohibition of discrimination based on disability, Congress passed the American's Disability Act of 1990.

In 1993, Felbatol (felbamate) and Neurontin (gabapentin) are FDA approved. The first new epilepsy drug treatment in more than 10 years received FDA approval July 29.

In 1994, Lamictal (lamotrigine) is FDA approved. The FDA has approved Lamictal Tablets (generic: lamotrigine) as an additional therapy for partial seizures in children aged two and up. This expands the already approved use in adults with partial seizures and for the generalized seizures of Lennox-Gastaut syndrome in children.

In 1996, the FDA approves Topamax (topiramate).

In 1997, FDA approves Gabitril (Tiagabine).

In 1997, The U.S. Food and Drug Administration (FDA) approved vagus nerve stimulation in combination with seizure medication for partial epilepsy in adults.

In 1999, FDA approves Keppra (levetiracetam). The FDA approved Keppra for use in adults as a secondary treatment for epilepsy in **1999**. A new study presented today at the Child Neurology Society meeting suggests that Keppra may also be effective as a first-line treatment for children who do not respond to traditional therapies.

In 2000, Trileptal (Oxcarbazepine) and Zonegran are approved by the FDA. Also in **2000**, A landmark conference, "Curing Epilepsy: The Promise and the Challenge," organized by the Epilepsy Foundation of America, sets confident goals for future treatment as well as prevention and cure of epilepsy; no seizures or side effects for those with the disorder; and discovering ways to stop epilepsy caused from injury, infection, or errors of development.

There is no other disorder or disease that has been given so many different names in the course of history as epilepsy. From this, we can conclude that throughout the ages people have been preoccupied with this disease. There are two main reasons for this interest:

First, epilepsy has always been a **common disorder**: zero, 5-1% of all the people who have been diagnosed with epilepsy.

SECTION 2: LEARNING HOW TO COPE WITH EPILESY

What is epilepsy? Does your life change when you develop the disorder? How do you cope with it? In Section 2, I explain what epilepsy is and how to live a happy productive life with the disorder.

EPILEPSY

The Most Important Secrets You Must Learn In Order To Live, Learn, and Be Happy With Epilepsy

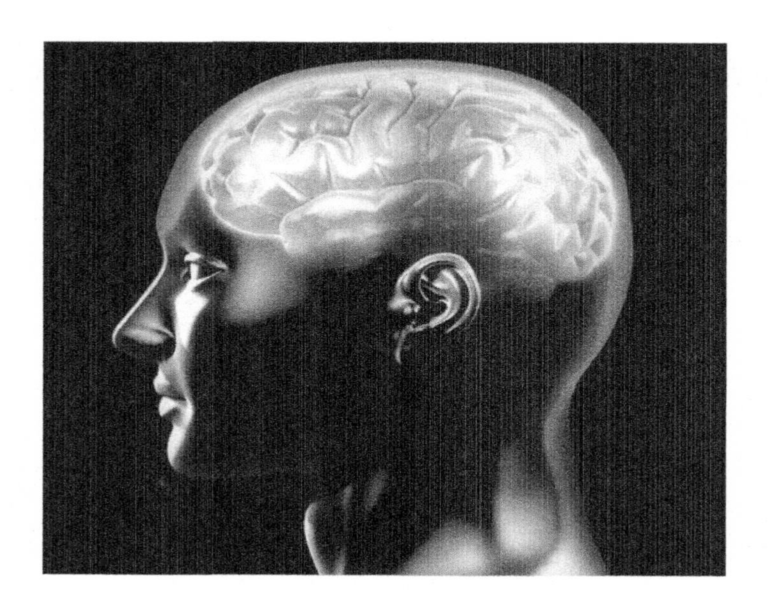

Chapter 2: What is Epilepsy?

Epilepsy affects millions of people worldwide, with more than two million people in the United States suffering from the disorder. Epilepsy is diagnosed in 125,000 Americans each year. It is estimated that epilepsy occurs ten times more frequently than multiple sclerosis and 100 times more often than the motornueron disease. Statistics show that one out of twenty people will have at least one epileptic seizure once in his or her life. One out of 200 will ultimately develop full-blown epilepsy. According to the Epilepsy Foundation of America, the causes of 70 % of all cases of epilepsy are unknown. Children and adolescents are more likely to have epilepsy of unknown or genetic origin.

Epilepsy is a neurological condition, which affects the nervous system. Another name for epilepsy is a *seizure disorder*. It is usually diagnosed after a person has had at least two seizures that were not caused by some known medical condition like alcohol withdrawal or extremely low blood sugar.

Brain injury or infection can cause epilepsy at any age. Studies done a short time ago showed that up to 70% of children and adults with recently diagnosed with epilepsy can be treated successfully. About 10% of children with epilepsy have seizures that do not react to treatment. The percentage of adults is higher, up to 15%. Up to 5% of the world's population may have a single seizure at some time in their lives. It is likely that around 50 million people in the world have epilepsy at any one time.

Epilepsy is a chronic disorder of cerebral function characterized by recurring convulsive seizures. Many conditions have epilepsy seizures. Unexpected discharge of additional electrical activity, which can be either generalized (involving many areas of cells in the brain) or focal, also known as partial (involving one area of cells in the brain), initiates the epilepsy seizure. Generalized seizures are called tonic-clonic (grand mal), in which there is loss of consciousness and spontaneous contraction of all the muscles of the body, lasting a few minutes; or absence (petit mal), in which there is clouding of the consciousness for about 1 to 30 seconds and no falling, with as many as 100 attacks occurring daily.

The Definition of Epilepsy

Epilepsy is when a person's brain cells frequently transmit information to the rest of the body by way of orderly electrochemical signals. These signals are not transmitted aimlessly; they do not course pell-mell through our nervous system. They are transmitted selectively, as some messages are repressed and others allowed long-term. This selectively prevents "cross talk" or message overload in the body's communication system.

Occasionally, a group of brain cells simultaneously "fires" or discharges a large number of electrical signals that produce a temporary rise in activity in certain parts of the brain, thus disrupting a person's internal communication system. This is a seizure. A seizure disturbs a person's consciousness, much in the way a lightning storm can disturb the electrical power supply. This disruptive overload of brain activity causes the strange body movements, unusual changes in speech, blank stare, and twitching of the eyes (clonic attack extremities) which are symptomatic of epileptic seizures.

A single seizure does not necessarily signal epilepsy. Epilepsy involves frequent seizures, varying from one or more a day, to one a month or even as few as one or two year. Seizures have many causes, epilepsy being only one of them. Having one or two seizures does not mean someone has epilepsy. Non-epileptic seizures can be caused by, among other things, high fevers, and alcohol withdrawal.

Epilepsy is a neurological condition, which affects the nervous system. Epilepsy is also known as a seizure disorder. It is usually diagnosed after a person has had at least two seizures that were not caused by some known medical condition like alcohol withdrawal or extremely low blood sugar.

The seizures in epilepsy may be related to a brain injury or a family tendency, but most of the time the cause is unknown. The word "epilepsy" does not indicate anything about the cause of the person's seizures, what type they are, or how severe they are.

A seizure is a sudden surge of electrical activity in the brain that usually affects how a person feels or acts for a short time. Seizures are not a disease in themselves. Instead, they are a symptom of many different disorders that can affect the brain. Some seizures are unnoticeable by others, while others are totally disabling.

The seizures in epilepsy may be related to a brain injury or a family tendency, but often the cause is completely unknown. The word "epilepsy" does not point out anything about the cause or seriousness of the person's seizures. About half of the people who have one seizure without an apparent cause will have another seizure, usually within 6 months. You are twice as likely to have another seizure if you have a known brain injury or other type of brain abnormality. If you do have two seizures, there is about an 80% chance that you will have more.

If your first seizure occurred at the time of an injury or infection in the brain, you are more likely to develop epilepsy than if you had not had a seizure in that situation. More seizures are also likely if your doctor finds abnormalities on a neurological examination; a set of tests of the functioning of your nervous system that is performed in the doctor's office.

Another thing that can help your doctor predict whether you will have more seizures is an EEG, electroencephalogram, a test in which wires attached to your scalp record your brain waves. Certain patterns on the EEG are typical of epilepsy. If your brain waves show patterns of that type, you are about twice as likely to develop epilepsy as someone who does not have those patterns.

A seizure is a sudden alteration of behavior due to a temporary change in the electrical functioning of the brain, in particular the outside rim of the brain called the cortex. Below you will find some of the symptoms people with epilepsy may experience before, during and after a seizure. Seizures can take on many different forms and seizures affect different people in different ways.

Epilepsy can develop in any person at any age. 0.5% to 2% of people will develop epilepsy during their lifetime. People with certain conditions may be at greater risk. Doctors for epilepsy have treated about 2.5 million Americans in the past five years. That is eight or nine out of every 1,000 people. In other words, out of 60,000 people filling a big stadium, about 500 have epilepsy. More men than women have epilepsy. New cases of epilepsy are most common among children, especially during the first year of life. The rate of new cases gradually declines until about age 10, and then becomes stable. After age 55 or 60, the rate starts to increase, as people develop strokes, brain tumors, or Alzheimer's disease. (These disorders can cause epilepsy.) There is a fine balance in the brain between factors that begin electrical activity. They are also factors that restrict it. There are systems that limit the spread of electrical activity. During a seizure, these limits break down, and abnormal electrical discharges can occur and spread to whole groups of neighboring cells at once. This linkage of electrical discharges creates a "storm" of electrical activity in the brain. This is a seizure. When a person has had at least two of these seizures, this is considered epilepsy.

The reasons why epilepsy begins are different for people of different ages. What is true for every age is that the cause is unknown for about half of everyone with epilepsy. Children may be born with a defect in the structure of their brain, or they may suffer a head injury or infection that causes their epilepsy. Severe head injury is the most common known cause in young adults. In middle age, strokes, tumors, and injuries are more frequent. In people over 65, stroke is the most common known cause, followed by degenerative conditions such as Alzheimer's disease.

Often seizures do not begin immediately after a person has an injury to the brain. Instead, a seizure may happen many months later. We do not have a good explanation for this common observation, but scientists are actively researching this subject.

Risk Factors for Developing Epilepsy

- ✓ Babies who are small for their gestational age
- ✓ Babies who have seizures in the first month of life
- ✓ Babies who are born with abnormal brain structures
- ✓ Bleeding into the brain
- ✓ Abnormal blood vessels in the brain
- ✓ Serious brain injury or lack of oxygen to the brain
- ✓ Brain tumors
- ✓ Infections of the brain: abscess, meningitis, or encephalitis
- ✓ Stroke resulting from blockage of arteries
- ✓ Cerebral palsy
- ✓ Mental handicap
- ✓ Seizures occurring within days after head injury ("early posttraumatic seizures")
- ✓ Family history of epilepsy or fever-related seizures
- ✓ Alzheimer's disease (late in the illness)

- ✓ Fever-related (febrile) seizures that are unusually long
- ✓ Use of illegal drugs such as cocaine

Mild head injuries, such as a concussion with just a very brief loss of consciousness, do not cause epilepsy. It may seem obvious that heredity (genetics) plays an important role in many cases of epilepsy in very young children, but it can be a factor for people of any age. For example, not everyone who has a serious head injury (a clear cause of seizures) will develop epilepsy. Those who do develop epilepsy are more likely to have a history of seizures in their family. This family history suggests that it is easier for them to develop epilepsy than for others with no genetic predisposition.

Epilepsy in which the seizures begin from both sides of the brain at the same time is called primary generalized epilepsy. Primary generalized epilepsy is more likely to involve genetic factors than partial epilepsy, in which the seizures arise from a limited area of the brain. Their risk is slightly higher than usual, not because they will "catch" it (that cannot happen) but because there may be a genetic tendency in the family that makes seizures and epilepsy more likely. Even so, most of them will not develop epilepsy. Epilepsy is more likely to occur in a brother or sister if the child with epilepsy has primary generalized seizures. Depending on the type of epilepsy and the number of family members who are affected, only about 4% to 10% of the other children in the family will have epilepsy.

Less than two people out of every 100 (2%) develop epilepsy at some point during their lifetime. The risk for children whose father has epilepsy is only slightly higher. If the mother has epilepsy and the father does not, the risk is still less than 5%. If both parents have epilepsy, the risk is a bit higher. Most children will not inherit epilepsy from a parent, but the chance of inheriting epilepsy is higher for some types.

If you have epilepsy, it is normal for you to be afraid that your children will have epilepsy too. However, a fear that your children will have epilepsy is not enough reason to decide against having any. The risk is low, most children outgrow epilepsy, and most people who have it are able to control their seizures by taking one medicine.

About 80% of people with epilepsy treated with seizure medicines remain free of seizures for at least 2 years. Many never have any more seizures. The chances of becoming completely seizure-free are best if there is no known brain injury or abnormality, and if the person has a normal neurological examination and EEG.

In adults, 50-60% will be seizure-free after using their first seizure medicine. Another 11-20% will gain seizure control using the second medication, leaving 20-30% who is still having seizures.

Among those who are young when their epilepsy is diagnosed, 20% are "smooth sailors"—they start on medication and never have another seizure after medication is stopped, even when they reach adulthood. About 50-60% of children become seizure-free with the first medication used, but 30% never stop taking seizure medicines. About 10% have a difficult time with "intractable seizures."

The more time that passes without seizures, the greater is the chance of staying seizure-free. Over 50% of children outgrow their epilepsy. Twenty years after the diagnosis, three-quarters of people will have been seizure-free for at least 5 years, although some may still need to take daily medication.

Many people who are seizure-free for 2 to 4 years can stop taking their medications, under their epileptologist supervision, without having further seizures. However, about 30% of children and 30% to 65% of adults will have seizures again. You need to discuss this with your neurologist and agree on a plan for stopping gradually over weeks or months, not all at once. Currently, most neurologists and epileptologists in the United States and Canada consider withdrawing seizure medicines after someone has been seizure-free for 1 to 2 years.

Some people with uncontrollable seizures with tolerable doses of seizure medicine do eventually become seizure-free. The longer that you continue to have seizures after the diagnosis of epilepsy is made, however, the lower the chance that your seizures will stop. Your epileptologist will want you to try different medications or combinations of them. The more medicines that fail to control your seizures, however, the less likely it is that another medication regimen will fully succeed. Other kinds of treatments, such as vagus nerve stimulation or epilepsy surgery, may be very helpful for some people who continue to have seizures while taking seizure medicines.

The life expectancy of people who have epilepsy is the same as for anybody else if they are otherwise healthy. Things like a stroke because some people's epilepsy or a brain tumor may die sooner from those conditions, of course.

A long-lasting convulsive seizure (called "tonic-clonic status epilepticus") is a medical emergency. If not stopped within about 30 minutes, it may cause permanent injury or death. In addition, people with epilepsy can also die from inhaling vomit during or just after a seizure. This can be prevented if someone will turn the person onto one side when the seizure begins and ensures that the vomit completely comes out of the mouth. Seizures are hardly ever fatal, even if the person loses consciousness.

People who are not seizure-free need to be careful about possible accidents during a seizure. Death from drowning is more common among people with epilepsy. It can even occur in a tub with only a few inches of water, so people who have seizures probably should stick to showers instead of baths. If you have epilepsy, your doctor— and the agency in your state or province responsible for licensing drivers— will help, you decide whether it is safe for you to drive. You should also be careful on train or subway platforms and when walking near busy streets. However, with some planning, you should be able to lead a life that is both active and safe.

Epilepsy has affected people since the creation of man. It has been acknowledged since the earliest medical writings. We now understand that epilepsy is a common disorder resulting from seizures that temporarily impair brain function. Few medical conditions have attracted so much attention and generated so much controversy. Throughout history, people with epilepsy and their families have suffered unfairly because of the ignorance of others. Fortunately, the stigma and fear generated by the words "seizures" and "epilepsy" have decreased during the past century, and most people with epilepsy now lead normal lives.

Epilepsy is a disorder of brain and nerve cell function that may or may not be associated with damage to brain structures. Brain *function* can be temporarily disturbed by many things, such as extreme fatigue; the use of sleeping pills, sedatives, or general anesthesia; or high fever or serious illness. "Brain damage" implies that something is permanently wrong with the brain's structure. This kind of damage may occur with severe head injury, cerebral palsy, or stroke or it may occur long before birth, with malformation or infection. Injuries to the brain are the cause of seizures in some people with epilepsy, but by no means all of them.

Brain injuries range from undetectable to disabling. Although brain cells usually do not regenerate, most people can make substantial recoveries. Brain damage, like epilepsy, carries a stigma, and some people may unjustly consider brain-injured patients "incompetent."

Many people mistakenly believe that people with epilepsy are also mentally handicapped. In the large majority of cases, this is not true. Like any other group of people, people with epilepsy have different intellectual abilities. Some are brilliant and some score below average on intelligence tests, but most are somewhere in the middle. They have normal intelligence and lead productive lives. There are some people, however, who have epilepsy associated with brain injuries that may also cause neurological impairments, including mental handicap. With only very rare exceptions, seizures do not cause mental handicap.

The belief that people with epilepsy are violent is an unfortunate image that is both wrong and destructive. People with epilepsy have no greater tendency toward severe irritability and aggressive behaviors than do other people.

Occasionally, problems associated with epilepsy, such as injury to specific brain areas or sensitivity to certain medications, can contribute to aggressive or confused behavior. Anxiety and depression may be slightly more common among people with epilepsy.

Single tonic-clonic seizures lasting less than 5-10 minutes are not known to cause brain damage or injury. However, there is evidence that more frequent and more prolonged tonic-clonic seizures may occur in some patients who injure their brain. Prolonged or repetitive complex partial seizures (a type of seizure that occurs in clusters without an intervening return of consciousness) also can potentially cause long-lasting impairment of brain function.

Some people have difficulty with memory and other intellectual functions after a seizure. These problems may be caused by the aftereffects of the seizure on the brain, by the effects of seizure medicines, or both. Usually, however, these problems do not mean that the brain has been damaged by the seizure. There may be a cumulative, negative effect of many tonic-clonic or complex partial seizures on brain function, but this effect appears to be rare. Most cases of epilepsy are not inherited, although some types are genetically transmitted (that is, passed on through the family). Most of these types are easily controlled with seizure medicines.

Generally, people with epilepsy have seizures and require medication for only a small portion of their lives. More than half of childhood forms of epilepsy are outgrown by adulthood. With many forms of epilepsy in children and adults, when the person has been free of seizures for 1 to 3 years, medications can often be slowly withdrawn and discontinued under a doctor's supervision.

Common Myths and Misunderstandings about Epilepsy

Even though medical progress has come a long way, the

myths and misconceptions of the past still pose a major problem.

<u>Some popular myths:</u>

Myth #1 - Epilepsy is contagious.

You cannot catch epilepsy from another person!

Myth #2 - You **can swallow your tongue during a seizure.**

It is physically impossible to swallow your tongue. In fact, you should never force something into the mouth of someone having a seizure. That is a good way to chip teeth, puncture gums, or even break someone's jaw.

Myth #3 - People with epilepsy are disabled and are unable to work.

People with the epilepsy have the same range of abilities and intelligence as anyone else. Some have severed seizures and cannot work; others are successful and productive in challenging careers. People with seizure disorders are found in all lifestyles and at all levels of business, government, the arts, and the professions.

Myth #4 - People with epilepsy look different.

Someone with epilepsy is actually having a seizure there is no way that his or her condition can be recognized.

Myth #5 - Epilepsy is a form of mental illness

Epilepsy is an umbrella term covering about twenty different types of seizure disorders. It is a functional, physical problem, not a mental one.

Myth #6- With today's medication, epilepsy is largely a solved problem

Epilepsy is a chronic medical problem that for many people can be successfully treated. Unfortunately, treatment does not work for everyone and there is a critical need for more research.

The truth is that epilepsy is a very common disorder. Epilepsy can happen to anybody at any time. In the vast majority of cases, epilepsy should not stop someone from living healthy, productive life. It is too often people's misconceptions about epilepsy create the disability, not epilepsy itself. Many features of seizures and their immediate aftereffects can be easily misunderstood as "crazy" or "violent" behavior.

Unfortunately, police officers and even medical personnel may confuse seizure-related behaviors with other problems. However, these behaviors merely represent semiconscious or confused actions resulting from the seizure. During seizures, some people may not respond to questions, may speak gibberish, undress, repeat a word or phrase, crumple important papers, or may appear frightened and scream. Some are confused immediately after a seizure, and if they are restrained or prevented from moving about, they can become agitated and combative. Some people are able to respond to questions and carry on a conversation fairly well, but several hours later, they cannot remember the conversation at all.

Epilepsy is perfectly compatible with a normal, happy, and full life. The person's quality of life, however, may be affected by the frequency and severity of the seizures, the effects of medications, reactions of onlookers to seizures, and other disorders that are often associated with or caused by epilepsy.

Some types of epilepsy are harder to control than other types of epilepsy. Living successfully with epilepsy requires a positive outlook, a supportive environment, and good medical care. Coping with the reaction of other people to the disorder can be the most difficult part of living with epilepsy.

Acquiring a positive outlook may be easier said than done, especially for those who have grown up with insecurity and fear. Instilling a strong sense of self-esteem in children is important. Many children with long-term, ongoing illnesses—not only epilepsy but also disorders such as asthma or diabetes—have low self-esteem. This may be caused in part by the reactions of others and in part by parental concern, which fosters dependence and insecurity. Children develop strong self-esteem and independence through praise for their accomplishments and emphasis on their potential abilities.

EPILEPSY

The Most Important Secrets
You Must Learn
In Order To Live, Learn, and Be Happy With Epilepsy

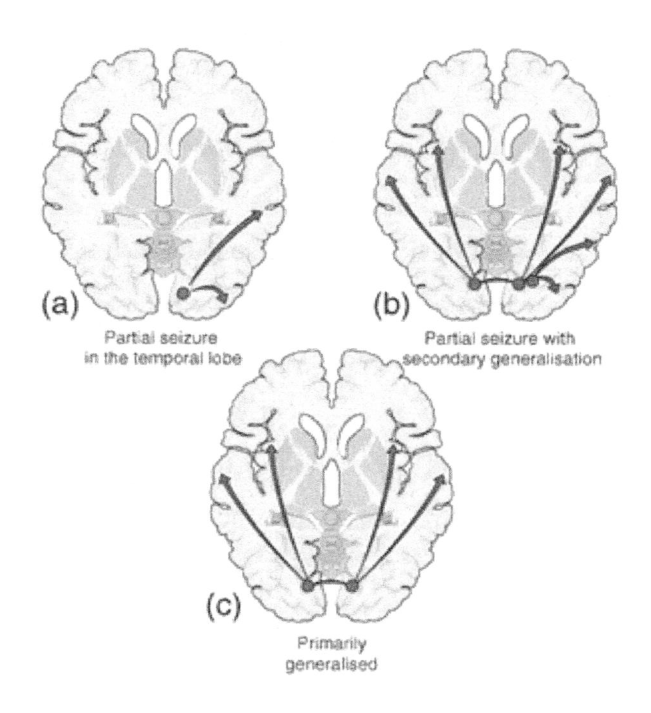

Chapter 3: The Brain and Epilepsy

Epilepsy is a disorder of the central nervous system, specifically the brain. In simple terms, our nervous system is a communications network that controls every thought, emotion, impression, memory, and movement, essentially defining who we are. Nerves throughout the body function like telephone lines, enabling the brain to communicate with every part of the body via electrical signals. In epilepsy, the brain's electrical rhythms have a tendency to become imbalanced, resulting in recurrent seizures.

If you have seen a picture of the brain before, it probably looked like this one, which illustrates the outer surface of the upper brain. This outer surface contains numerous folds that increase the surface area and allow more cerebral cortex to be packed into the skull, giving us more "brain power."

The brain is an extraordinarily complex organ. When it comes to understanding epilepsy, there are several concepts about the brain you will need to learn. The first is that the brain works on electricity. Normally, the brain continuously generates tiny electrical impulses in an orderly pattern. These impulses travel along the network of nerve cells, called neurons, in the brain and throughout the whole body via chemical messengers called neurotransmitters.

A seizure occurs when the brain's nerve cells misfire and generate a sudden, uncontrolled surge of electrical activity in the brain. Another concept important to epilepsy is that different areas of the brain control different functions.

If seizures arise from a specific area of the brain, then the initial symptoms of the seizure often reflect the functions of that area. The right half of the brain controls the left side of the body, and the left half of the brain controls the right side of the body. If a seizure starts from the right side of the brain, in the area that controls movement in the thumb, then the seizure may begin with jerking of the left thumb or hand.

The Brain

The upper brain, or cerebrum, is divided into left and right halves, called cerebral hemispheres. A bundle of nerve fibers called the corpus callosum connects these. Each cerebral hemisphere contains four lobes: frontal, parietal, occipital, and temporal. Each lobe contains many different areas that control a variety of functions.

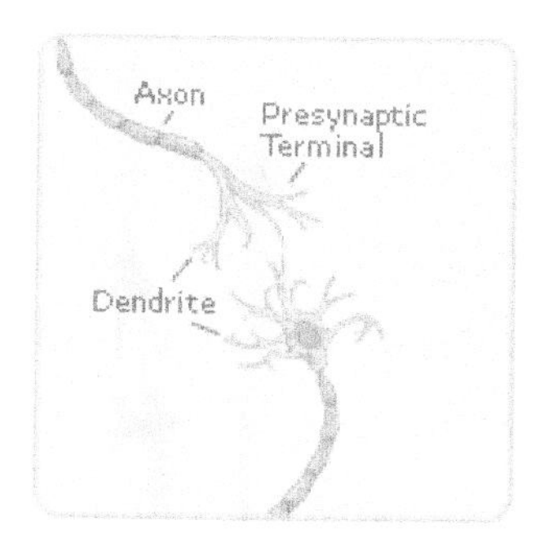

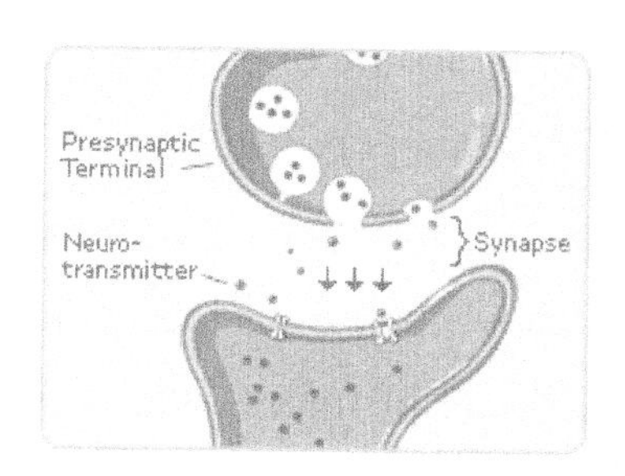

The brainstem and spinal cord

The lower part of the brain contains the brainstem, which controls sleep-wake cycles, breathing, and heartbeat. The upper part of the brainstem contains the thalamus and hypothalamus. The lower part of the brainstem is continued as the spinal cord, which carries messages between the brain and the rest of the body.

Nerve cells of the brain

Nerve cells, or neurons, are the building blocks of the brain. They work like computer chips, analyzing and processing information and then sending signals through the nerve fibers.

The nerve fibers act like telephone wires, connecting different areas of the brain, spinal cord, muscles, and glands.

Nerve cells are so small that a microscope is necessary in order to see them. There are approximately 100 billion nerve cells in the brain. During a seizure, each cell may fire as many as 500 times a second, much faster than the normal rate of about 80 times a second in the brain and spinal cord. Hundreds of impulses bombard each call every second. One of the wonders of the human brain is how the billions of individual computers (neurons) in the brain function in a coordinated fashion to control our movements and breathing, and most importantly, to allow us to think and feel.

Neurotransmitters

Neurotransmitters are the chemical messengers of the brain. These substances are released at the end of the cell and cross the synapse, a tiny space between the walls of one cell's axon and the dendrite of the next nerve cell, to bind to receptors located on that dendrite.

There are many kinds of neurotransmitters, but each individual nerve cell produces only one major type. Some of the neurotransmitters are carried a long distance within the nervous system. Others, however, have local effects; that is, they are produced by and released onto cells that are close to each other.

Neurotransmitters are important in diseases of the nervous system. In Parkinson's disease, for example, cells that manufacture dopamine, an important neurotransmitter that regulates movement, are lost. Loss of nerve cells may contribute to the development of epilepsy in some cases. For example, prolonged lack of oxygen may cause a selective loss of cells in the hippocampus, which may lead to epilepsy. Some of the major neurotransmitters in the brain shut off or decrease brain electrical activity. They cause nerve cells to stop firing. These neurotransmitters are called "inhibitory" because they inhibit the activity of the cells. A neurotransmitter called GABA is the best-known example of this type.

Other neurotransmitters stimulate or increase brain electrical activity. That is, they cause nerve cells to fire. These are described as "excitatory." Glutamate is an example of this type. According to one theory, epilepsy is caused by an imbalance between excitatory and inhibitory neurotransmitters. If the inhibitory neurotransmitters in your brain are not active enough, or if the excitatory ones are too active, you are more likely to have seizures.

Many of the new medicines being developed to treat epilepsy try to influence these neurotransmitters. They try to increase the activity of the inhibitory ones, which turn cells off, or reduce the activity of the excitatory ones, which turn cells on. Either way, the idea is to have less uncontrolled electrical activity in your brain, and therefore fewer seizures.

Chapter 4: When Your Child Develops Epilepsy

Watching your child have a seizure for the first time was probably one of the most frightening moments of your life. My parents told me that is most terrifying experiences of their life. Their bedroom was next to mine. One night they heard some funny noise coming from my room. They both went into my room to find me with my eyes rolled back twitching, lips blue, mouth chattering and body shaking.

Finding out that your child has epilepsy is a painful experience. When you have a child, you only want the best for your child. When your child gets a cold or a fever your stomach drops and you feel their pain. Can you imagine if you find out your child has epilepsy?

There may be overwhelming feelings of sadness or depression, as parents grieve for what they perceive as the loss of their healthy child. They grieve for the life changes that will their child will experience having epilepsy and they worry that their child will not be able to accomplish their dreams that lies in their child's future destiny. Feelings of grief are typical.

Many parents as ask themselves "Why did my child develop epilepsy?" Some parents blame themselves and feel that it is somehow their fault or wonder if they could have done something differently to prevent the seizures their child is currently experiencing. Let me tell you now that you cannot blame yourself. Your child developing epilepsy is not your fault. There are so many causes for developing epilepsy that it is impossible to pin point. Parents may feel resentful about the new challenges and demands they will be facing, and then experience guilt for feeling this way. These feeling are completely normal and all parents go through these emotions. Anxiety is the most common feeling because parents will have a number of overwhelming worries. **Some of them are:**

- ✓ Will my child die?
- ✓ Will he be brain damaged?
- ✓ Will my child experience problems with developmental skills?
- ✓ Can I let children play by themselves? Do I have to supervise them all the time?
- ✓ Can I send my child to after school activities?
- ✓ How do I explain to my child that they have epilepsy? What do I say to him or her?

Along with these worries, parents also struggle with the fact they do not know when their child's next seizure will occur. You will probably feel angry, scared, and resentful that you have no control over what will happen. You may worry about whether your child is safe while away from home, that teachers or other adults will mishandle seizures, or make the situation worse by overacting. Parents will also worry about the potential side effects of their child's medications or the impact of missing school due to seizures and medical appointments.

The future may suddenly seem terrifying and uncertain for both your child and your whole family. Your child having epilepsy is not as bad as it sounds. Medical research has found that most children who have a seizure do not have another one. The medical community has also found out that a majority of children who have epilepsy (which means that they have had more than one seizure) will outgrow their seizure disorder. Mostly all children with epilepsy are perfectly healthy and normal. Their intelligence level is not affected. They can participate in after school activities and any other activities that your child may enjoy.

Currently, medical research tells us that 70 to 80 percent of children can control their seizures with medication. There is no a cure for epilepsy. Many doctors feel medication is the answer for children. Many doctors say all children need is medication because their seizures not as severe as adults are and with medication, the seizures may go away on their own.

Once your child experiences seizures or epilepsy, it will probably change your family for life. A parent who has a child with epilepsy will have to obtain new responsibilities. You will need to make sure that your child is getting good medical care. You need to make sure that your child is seeing a doctor who knows a lot about epilepsy and is up to date with all the recent medical research and medications. The best doctors to see are epileptologist. These doctors focus only on the study of epilepsy. You can find a good epileptologist in your area by contacting an epilepsy clinic or your states epilepsy foundation. The main epilepsy foundation is in Maryland. They could probably direct you to the people you need to speak with.

You will have to make sure that your child takes their medication. Skipping a dose can cause a seizure. You can buy weekly medication organizers, so your child does not miss a dosage. You should do your own research and find out as much as you can about epilepsy. Do not rely just on the doctors. Doctors are human. They can make mistakes too.

You need to find out as much as you can about epilepsy and make sure you, your family and your doctors are doing everything possible to help your child. You may also have to become an advocate for your child, explaining epilepsy to family, friends, and teachers (you will have to inform the school. In case, your child has a seizure in school).

It is not going to be easy in the beginning. You will probably feel overwhelmed, stressed, and extremely upset. While it may be tough being the parent of a child with epilepsy sometimes, just remember that treatment works. Epilepsy is not nearly as scary as it sounds. Using the medical treatment for epilepsy will help your child, so they can live a healthy, productive life with few limitations.

A child developing epilepsy is a shock to everyone. It places new demands on the family. When one member of the family has epilepsy, everyone in the family is affected by the disorder. There are different ways to help cope with your child's disorder read- find out as much as you can about epilepsy: **such as:**

- ✓ What is epilepsy?
- ✓ What medication is available?

- ✓ What medication is use most for children?
- ✓ What are the side effects?
- ✓ What tests should I take my child take?
- ✓ Keep a journal
- ✓ Figure out how many seizures your child is having monthly
- ✓ Write done what do they do during their seizures
- ✓ Report each seizure to your doctor
- ✓ Find out the lasted medical techniques available
- ✓ Join support groups
- ✓ It may be hard at first to reach out for help, but talk to other parents who have gone through similar experiences. This helps you feel that you are not alone and that you have other parents that understand what you are going through.

- ✓ Do not be in denial. Accept that your child has epilepsy. Pretending that your child does not have epilepsy will only make matters worse. Sometimes, one parent will cope with their child's epilepsy by learning everything there is to know on epilepsy, while the other parent remains motionless. Holding all their emotions inside and making believe nothing is wrong.

Find out about the ketogenic diet - The ketogenic diet can help control some children with epilepsy and stop seizures without any medication. It works very well in many children. It is strict diet and difficult to follow. In fact, it is so difficult to follow that most doctors recommend it only for children who have not been able to control their seizures. The ketogenic diet is an extremely high-fat diet. Your child would have to eat four times as many fat calories as calories from protein or carbohydrate. A meal might include a small portion of chicken, a little bit of fruit, and a lot of fat, mostly butter or cream.

Your child may start the diet in a hospital, so nurses and doctors can observe the first few days. Your child will probably need to go without any food for 36 to 48 hours before beginning the diet. After that, your child's food is increased slowly over a few days. This diet does not provide all the vitamins a body needs, so your child will probably have to take sugar-free vitamin supplements. This diet is not for everyone. This is something you would have to discuss with your doctor. This diet is not for everyone.

Every parent goes through a period (It is usually right after your child develops epilepsy) that you want to find someone to blame when your child is diagnosed with epilepsy. Some people blame themselves or someone around them that they see frequently to blame. This is very common.

Every parent has different ways of dealing with their child's disorder. Eventually, they will learn to accept they child's disorder. However long it takes, most parents stop asking why and move forward. You need to realize we cannot change the past. The past is gone. The present is now and what we do in the present will affect our child's future. We need to help your child adjust and get on with their lives.

The best ways of dealing with the challenges of having a child with epilepsy is keeping the lines of communication open. Listen to your child. Show your child that you are interested in what he has to say. You do not need to understand, relate, or like what they are saying. Just listen.

- ✓ Speak to your spouse and close relatives about how you are feeling
- ✓ Talk to your children about epilepsy and seizures
- ✓ Take their concerns seriously and get outside help if necessary

- ✓ Call the Epilepsy Foundation and tell them about your child's concerns and yours and ask them to direct them to the people that can help
- ✓ Let your child know that it is okay to feel frustrated or angry, and that he or she can talk to you about their feelings.
- ✓ Do not let them feel different. Make them feel like they are just like everybody else
- ✓ How you deal with their disorder now will affect how they grow up

Children go through many similar feelings as their parents. When I was a child, I use to cry after a seizure because I wanted them to stop. Depending on the age of the child and. the type of seizures they are having, they may experience fear, anger, worry, or an overwhelming amount of emotion. They may feel afraid of having a seizure in public, in front of their friends or in school. They may feel confused or anxious about having to attend medical appointments, undergo tests, take medications, or miss school or other activities when a seizure has occurred. Children may feel angry if they have an aura (a feeling before a seizure.) They may get upset if they feel epilepsy is preventing them from taking part in activities with their friends.

Some children with epilepsy feel very isolated or different from their friends; being accepted by your friends is a crucial time in their development. It may be difficult in making close friends or they may not want to tell their friends about their seizures.

It is good to explain to your child that if someone wants to be your good friend that they are going to like you for whom you are and they are going to like everything about you. If they do not then they are not a true friend and you do not want them as a friend.

Usually children with epilepsy develop low self-esteem because they feel different or not as good as the other kids do. It may become worse if schoolmates think of them as weak or if they feel, overprotected by their families. It is very important to make sure your child feels good about them and has high self-esteem. I cannot even begin to stress the importance of having high self-esteem; it is the key to having a healthy, productive life for your child.

Make sure children loves themselves and accepts themselves and their disorder. Make your child understand that they need to be grateful of what they have and others may have it worse. Give them some examples. You need to help them let go of all those angry emotions inside. If your child holds anger inside it will not help them, it will only hurt them. You need to show them how to think positively and focus on their accomplishments. You need to help them focus on the goals and dreams. You need to help your child believe in itself. Help them develop a sense of pride in themselves. Tell them it does not matter what others think about them, what matters is how you think about yourself.

Tell your child not to feel sorry for them. No one is perfect. Studies have shown that people who have negative attitudes are more like to live chaotic lives. Many people have a hard time focusing on the positive because they allow their negative sides to consume them... I firmly believe that focusing on the negatives will causes seizures. When you help, your child feels a sense of accomplishment and self-worth. You are helping your child to overlook the negative. Confidence comes from our self-esteem. To have high self-esteem you need to help your child develop simple short and long term goals. You need to help your child create direction in their life. This will help them developed a greater sense of pride, inner strength and self-worth in themselves.

When your child has fears about seizures should address them right away and your child should feel that he or she could talk to mom or dad about any questions, concerns, or worries. It is important to not to focus completely on your child's disorder and focus more on other positive characteristics of your child. Most children with epilepsy have the same concerns, desires, and dreams as other children. They want to be like everyone else. They do not want to be treated any differently. Most important do not forget to tell your child that you love them and how special they are to you. Hearing those words from mommy or daddy means everything to them. It gives them the strength to know that mommy and daddy loves them just the way they are can help a child and give that child the strength and ambition to lead a productive and wonderful life.

Tell your child that they brighten your day. Let your child know how much joy they bring into your life. Show them how special they are to you and make sure you point out all their good qualities. The words I Love You, helps your child know that you love them unconditionally. You will love them even though they have epilepsy. Explain to them that their seizures make them no different from anyone else. They want you to love them no matter what.

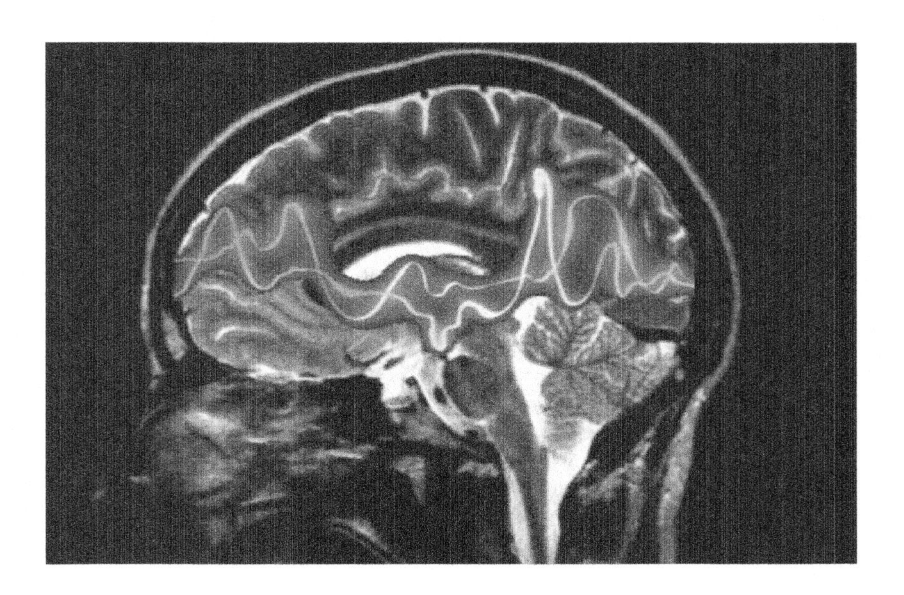

Chapter 5: The Different Types of Seizures Caused By Epilepsy

There are so many kinds of seizures that epileptologists who specialize in epilepsy are still updating their thoughts about how to group them. Typically, they classify seizures into two types, primary generalized seizures and partial seizures. The difference between these types is in how they begin. The first type of seizure I would like to discuss is an aura. An aura is considered a warning sign for many people who have seizures. I always experience an aura before I have a full-blown seizure.

An aura varies significantly between different people. Some people may experience an aura right before a seizure or several minutes to hours earlier. Usual warning signs right before seizures are changes in physical sensations, changes in your ability to interact with things happening outside you, and changes in how familiar the outside world seems to you. Additional warning signs that may happen hours before a seizure are depression, irritability, sleep disruption, queasiness, and headache.

People with complex partial seizures are the most likely to experience warning signs. Approximately 55% to 65% of people with these seizures experience some type of aura. It is unclear whether having seizures that arise from one particular side of the brain makes you more likely to have auras than people whose seizures arise on the other side.

Researchers who have studied numerous types of auras for many years have found patterns. For many people, the sensations in an aura occur in the same order.

For example:

1. You may always feel fear, then déjà vu (the illusion of remembering scenes and events when experienced for the first time)

2. You may have a feeling that one has seen or heard something before

3. You may experience something overly or unpleasantly familiar), then a strange taste in your mouth. This pattern would point towards the existence of a prominent seizure pathway in your brain.

The part of your brain where your seizures begins to formulate also may be associated to a specific type of aura because an aura represents the beginning of a seizure. Since different parts of the brain are responsible for different things, the warning signs you experience will be related to the functions of the section of brain where the seizure is about to occur. People whose seizures begin in the temporal lobe tend to have certain types of auras and those whose seizures begin somewhere else often will have several types.

Some people have a distinctive feeling or some other warning sign when a seizure is coming. A warning like this is also called an aura. Although some auras are unpleasant, they can be helpful because they can give you time to prepare for the seizure and keep yourself from being injured. Most injuries from seizures happen if there is no warning sign, if the warning is not recognized, or if there is not enough time to react to it.

In 1987, Taylor and Lochery proposed three basic categories for auras: simple primitive, special-senses, and intellectual. Anxious or nervous feelings, epigastric sensations, fear, and the beyond description the feeling of a seizure coming on are considered simple prehistoric auras. These are considered "primitive" because they deal with basic emotions, primal sensations, and ancient instincts. Epigastric sensations **(odd feelings in the stomach)** are the most common auras of this kind. Simple primitive auras tend to occur alone, not as part of a sequence of sensations or thoughts, except that they may be associated with anxiety.

Some doctors have speculated that simple primitive auras are usually associated with epilepsy or seizures originating in the temporal lobe of the brain. Mesial temporal sclerosis **(a scarring of tissue usually involved in temporal lobe seizures)** has been associated with the experience of simple primitive phenomena. Others have found that people who experience fear auras almost always have seizures that start in the front or middle part of the temporal lobe.

Some researchers have noticed that simple primitive auras occur mainly in patients with temporal lobe epilepsy that began early in life. The largest part of the brain, the cerebrum, is divided into four paired sections, called lobes-the frontal, parietal, occipital and temporal lobes. Each lobe controls a specific group of activities. The temporal lobe, located on either side of the brain just above the ear, plays an important role in hearing, language, and memory. In people with temporal lobe epilepsy, the area where the seizures start - called the seizure focus - is located within the temporal lobe. This is the most common type of epilepsy in teens and adults.

Auras that involve "special senses" include seeing, hearing, or tasting things that most people would not see, hear, or taste. A sense of the room closing in or spreading out, a strange noise or a bitter, harsh taste can all represent an aura of this kind. Less ordinary examples are feelings of as if things are not real or out-of-body experiences. Sudden changes in mood other than fear and anxiety are considered to belong in this group.

Perhaps the most commonly experienced special-senses aura is déjà vu. Feelings of déjà vu are among the most common of all auras. Auras in this group have not been closely studied. One group of researchers has proposed that psychic auras like déjà vu and Autonomic auras like epigastric sensations are more common in patients with right-sided EEG abnormalities. Much more information is needed before any conclusions can be drawn.

Intellectual auras have to do with higher-order functions such as language and complex thought. Language disorders during auras include word-finding difficulty, impaired comprehension during a conversation or substituting another word for the desired word (for instance, saying "chicken" instead of "children"). Auras of thinking include crowding or acceleration of thoughts, forced thinking (not being able to control your thinking, like getting one word or phrase stuck in your head), and other changes in quality of thought.

One study found that intellectual auras happened most frequently in males with high IQs. Again, more research is necessary to draw any definite conclusions. Studies have shown that no specific aura is linked with a poor or negative outcome. Some researchers have found that fear during a seizure is associated with increased psychological and behavioral disturbance, but the reasons for this are unclear. Take advantage of the opportunity to look for a safe place to experience the seizure that is likely to follow. Telling other people about your warning signs will make it easier for them to help you and can help them to understand epilepsy. You may be able to say, "I'm not feeling well" or "I'm going to have a seizure." On the other hand, you may just do something that other people or even animals, such as some dogs can recognize: You may suddenly get a strange expression on your face. You may stop talking in the middle of a sentence. If you are listening to someone else, you may seem to drift off and not pay attention. Your whole body may tense up. Family members and friends who recognize the meaning of these signs can help you stay safe during your seizure. Auras are not something you are able to control-they are caused by some small part of your brain that is misfiring. As strange as your aura may be, your doctor has probably heard of some that are even stranger.

It is important to tell your doctor about your auras so that your diagnosis will be more accurate and your treatment will be more focused. Knowing what type of auras you have can help your doctor find where your seizures begin. Although auras alone are rarely used to make a diagnosis, they can be used to confirm one. Call your doctor if you experience any warning signs on a regular basis, no matter how often they happen. If you experience a familiar sign that is not followed by any other seizure activity, make a note of it and tell your doctor about it anyway. You may be having small seizures, or another related condition that has been untreated. It is a good idea to keep track of your auras and seizures on a calendar so your report to the doctor will be accurate.

Take many forms. Before your doctor can prescribe the right treatment, he or she must figure out which type (or types) you have. That's the purpose of all the tests discussed in the diagnosis section—not just to tell whether you have epilepsy but also to tell what kind.

Primary generalized seizures begin with a widespread electrical discharge that involves both sides of the brain at once. Hereditary factors are important in many of these seizures. Partial seizures begin with an electrical discharge in one limited area of the brain. Some are related to head injury, brain infection, stroke, or tumor, but in most cases, the cause is unknown. Partial seizures include Jacksonian epilepsy, characterized by jerking in the hand and face on the side opposite the brain activity; and psychomotor seizures, in which there may be localized convulsion with no loss of consciousness, as well as incoherent speech and various involuntary movements of the body. Often a warning cluster of signs and symptoms called an aura accompanies these. The cause is unknown in over half the cases of epilepsy, especially in those with onset under age 20. Predisposing factors in other cases include familial history, head injury, alcohol withdrawal, infections (such as meningitis), and abnormalities (such as tumors) of the brain. One question that is used to further classify partial seizures is whether consciousness (the ability to respond and remember) is "impaired" or "preserved." The difference may seem obvious, but really, there are many degrees of impairment or preservation of consciousness.

Identifying certain seizure types and other characteristics of a person's epilepsy like the age at which it begins, for instance, allows doctors to classify some cases into epilepsy syndromes. This kind of classification helps the medical community know how long the epilepsy will last and what is the best way to treat it.

Primary Generalized Seizures

- ✓ Absence seizures
- ✓ Atypical absence seizures
- ✓ Myoclonic seizures
- ✓ Atonic seizures
- ✓ Tonic seizures
- ✓ Clonic seizures
- ✓ Tonic-clonic seizures

Partial Seizures

- ✓ Simple partial seizures
- ✓ Complex partial seizures
- ✓ Secondarily generalized seizures

Absence seizures cause a short loss of consciousness (just for a few seconds) with few or no symptoms. The patient, most often a child, typically interrupts an activity and stares blankly. These seizures begin and end abruptly and may occur several times a day. Patients are usually not aware that they are having a seizure, except that they may be aware of "losing time." Absense seizures are usually less than 10 seconds, but it can be as long as 20. They begin and end suddenly. Absence seizures are brief episodes of staring. Another name for them is petit mal. During the seizure, awareness and responsiveness are impaired. People who have them usually do not realize when they have had one. There is no warning before a seizure, and the person is completely alert immediately afterward.

Simple absence seizures are just stares. Many absence seizures are considered *a complex absence seizure, which means* that they include a change in muscle activity. The most common movements are eye blinks. Other movements include slight tasting movements of the mouth, hand movements such as rubbing the fingers together, and contraction or relaxation of the muscles. Complex absence seizures are often more than 10 seconds long. Absence seizures usually begin between ages 4 and 14. The children who get them usually have normal development and intelligence. In nearly 70% of cases, absence seizures stop by age 18. Children who develop absence seizures before age 9 are much more likely to outgrow them than children whose absence seizures start after age ten years old. Children with absence seizures do have higher rates of behavioral, educational, and social problems. Absence seizures can resemble some complex partial seizures or episodes of daydreaming:

Atypical seizures are similar to typical seizures accept that they tend to begin more slowly, last longer (up to a few minutes), and can include slumping or falling down. The person may also feel confused for a short time after regaining consciousness. While the cause of atypical seizures may be unidentifiable, they are sometimes traced to abnormalities in the brain that were present at birth (congenital) or from trauma or injury, or from complications from liver or kidney disease. Atypical means unusual or not typical. The person will stare (as they would in any absence seizure) but often is somewhat responsive. Eye blinking or slight jerking movements of the lips may occur. This behavior can be hard to distinguish from the person's usual behavior, especially in those with lower intelligence. Unlike other absence seizures, rapid breathing usually cannot produce these seizures. They generally begin before age 6. Most of the children affected have below-average intelligence and other types of seizures that are difficult to control. Many have Lennox-Gastaut Syndrome. An atypical absence seizure usually continues into adulthood and sometimes-ordinary behavior for these children will look like an atypical absence seizure. Daydreaming and inattentiveness can mimic these seizures

Myoclonic seizures consist of sporadic jerks, usually on both sides of the body. Patients sometimes describe the jerks as brief electrical shocks. When violent, these seizures may result in dropping or involuntarily throwing objects.

The diagnosis can be difficult if the behavior during seizures is similar to the child's usual behavior. The EEG (electroencephalogram), which records brain waves, will be used, but most children with these seizures have patterns on their EEG when they're not having a seizure that are similar to the seizure pattern. They are very brief jerks. Usually they do not last more than a second or two. There can be just one, but sometimes many will occur within a short time. Myoclonic Seizures are brief, shock-like jerks of a muscle or a group of muscles. "Myo" means muscle and "clonus" means rapidly alternating contraction and relaxation—jerking or twitching—of a muscle.

Even people without epilepsy can experience myoclonus in hiccups or in a sudden jerk that may wake you up as you is just falling asleep. These things are normal. In epilepsy, myoclonic seizures usually cause abnormal movements on both sides of the body at the same time. They occur in a variety of epilepsy syndromes that have different characteristics:

The seizures usually involve the neck, shoulders, and upper arms. In many patients, the seizures most often occur soon after waking up. They usually begin around puberty or sometimes in early adulthood in people with a normal range of intelligence. In most cases, these seizures can be controlled with medication but it must be continued throughout life.

Lennox-Gastaut Syndrome (LGS) is a rare disorder that typically becomes apparent during infancy or early childhood. The disorder is characterized by frequent episodes of uncontrolled electrical disturbances in the brain (seizures) and, in many cases, abnormal delays in the acquisition of skills that require the coordination of mental and muscular activity (psychomotor retardation). Individuals with the disorder may experience several different types of seizures. Lennox-Gastaut Syndrome may be due to or occur in association with a number of different underlying disorders or conditions.

Progressive Myoclonic epilepsy has rare syndromes. In this category features a combination of myoclonic seizures and tonic-clonic seizures. Treatment is usually not successful for very long, as the patient deteriorates over time. The epileptic syndromes that most commonly include myoclonic seizures usually begin in childhood, but the seizures can occur at any age. Other characteristics depend on the specific syndrome.

The outlook for patients with the various syndromes that include myoclonic seizures varies widely. See the specific syndromes for more information. As mentioned, some episodes of myoclonus are normal. Some myoclonic seizures occur in reflex epilepsies, triggered by flashing lights or other things in the environment.

Atonic seizures are easy to identify. The syndromes usually can be diagnosed based on the medical history and often EEG patterns. Atonic seizures last less than 15 seconds. Muscle "tone" is the muscle's normal tension. "Atonic" means "without tone," so in an atonic seizure, muscles suddenly lose strength. The eyelids may droop, the head may nod, and the person may drop things and often falls to the ground. These seizures are also called "drop attacks" or "drop seizures." The person usually remains conscious. Another name for this type of seizure is "akinetic," which means "without movement." Atonic seizures often begin in childhood. They often last into adulthood. Many people with atonic seizures are injured when they fall, so they may choose to use protection such as a helmet.

Patients who have seizures that cause them to fall when they are standing often have tonic seizures (involving sudden muscle contraction) rather than atonic seizures. Usually descriptions of the seizures by witnesses will suggest the diagnosis. Some EEG monitoring may be performed to confirm it. If the seizures persist, other tests may be used to make sure that changes in the heart rhythm or blood pressure are not causing the patient to fall down.

Partial seizures are divided into simple, complex and those that evolve into secondary generalized seizures. The difference between simple and complex seizures is that during simple partial seizures, patients retain awareness; during complex partial seizures, they lose awareness.

Simple partial seizures are further subdivided into four categories according to the nature of their symptoms: motor, autonomic, sensory, or psychological. Motor symptoms include movements such as jerking and stiffening. Sensory symptoms caused by seizures involve unusual sensations affecting any of the five senses (vision, hearing, smell, taste, or touch). When simple partial seizures cause sensory symptoms only (and not motor symptoms), they are called "auras."

Autonomic symptoms affect the autonomic nervous system, which is the group of nerves that control the functions of our organs, like the heart, stomach, bladder, intestines. Therefore, autonomic symptoms are things like racing heartbeat, stomach upset, diarrhea, loss of bladder control. The only common autonomic symptom is a peculiar sensation in the stomach that is experienced by some patients with a type of epilepsy called temporal lobe epilepsy. Simple partial seizures with psychological symptoms are characterized by various experiences involving memory (the sensation of déjà-vu), emotions (such as fear or pleasure), or other complex psychological phenomena.

Complex partial seizures mean impairment of awareness. Patients seem to be "out of touch," "out of it" or "staring into space" during these seizures. There may also be some "complex" symptoms called automatisms. Automatisms consist of involuntary but coordinated movements that tend to be purposeless and repetitive. Common automatisms include lip smacking, chewing, fidgeting, and walking. The third kind of partial seizure is one that begins as a focal seizure and evolves into a generalized convulsive ("grand-mal") seizure.

Most patients with partial seizures have simple partial, complex partial and secondarily generalized seizures. In about two-thirds of patients with partial epilepsy, seizures can be controlled with medications. Partial seizures that are unable to be treated with medications can often be treated surgically.

There are six types of **generalized seizures**. The most common and dramatic, and therefore the most well known, is the generalized convulsion, also called the grand-mal seizure. In this type of seizure, the patient loses consciousness and usually collapses. The loss of consciousness is followed by generalized body stiffening (called the "tonic" phase of the seizure) for 30 to 60 seconds, then by violent jerking (the "clonic" phase) for 30 to 60 seconds, after which the patient goes into a deep sleep (the "postictal" or after-seizure phase). During grand-mal seizures, injuries and accidents may occur, such as tongue biting and urinary incontinence.

Chapter 6: Epilepsy Medications and Treatments

Medications to prevent epilepsy seizures are called antiepileptic. The goal is to find an effective antiepileptic medication that causes the fewest side effects.

Antiepileptic medications prevent seizures in 60% to 70% of people who take them. Although many people experience side effects, medication is still the best way to prevent epileptic seizures. The benefits of treatment with medication usually outweigh the drawbacks.

There are many antiepileptic medications (called AEDs, anticonvulsants, or ant seizure drugs), but they do not all treat the same types of seizures. The first step in choosing a medication to treat your seizures is to identify the types of seizures you have. Many drugs are available to treat epilepsy, several of which have only recently been released.

Older, classic medications used to treat epilepsy include:

- ✓ Phenytoin (Dilantin or Phenytek)
- ✓ Phenobarbital
- ✓ Carbamazepine (Tegretol or Carbatrol)
- ✓ Primidone (Mysoline)
- ✓ Ethosuximide (Zarontin)
- ✓ Valproic acid (Depakene)
- ✓ Divalproex (depakote, Depakote ER)

- ✓ Diazepam (Valium) and various chemical derivatives such as Klonopin, Tranxene

Newer drugs to treat epilepsy include:

- ✓ Felbamate (Felbatol)
- ✓ Gabapentin (Neurontin)
- ✓ Lamotrigine (Lamictal)
- ✓ Tiagabine (Gabitril)
- ✓ Topiramate (Topamax)
- ✓ Levetiracetam (Keppra)
- ✓ Zonisamide (Zonegran)

Currently, more drugs are being studied to help control seizures.

In general, for a given type of epilepsy there are only minor differences among appropriate drugs. The choice is usually based on other factors specific to each patient, such as which side effects the patient can tolerate and which delivery method is acceptable. Although the different types of epilepsy vary greatly, in general, medications can control seizures in about

70% of epilepsy patients.

Monotherapy refers to treatment with a single drug. One of the main principles of drug therapy in epilepsy is to use one drug in sufficient dosages to achieve its maximum potential. (The most common treatment mistake is to treat a patient with two drugs, each at moderate dosages. This is usually not effective and may cause unwanted side effects.) Only when one drug proves to be ineffective should another drug or a drug combination be tried. In the United States, all of the recently released drugs that are currently approved to be used in combination with a "classic" drug, if necessary, when seizures do not respond to monotherapy. Certain combinations work better than others do, but combinations of more than two drugs should be avoided.

Side effects

Before any drug is prescribed, your health care provider will discuss with you the potential benefits, side effects, and risks. As is true of all drugs, the drugs used to treat epilepsy have side effects. The occurrence of side effects depends on the dose, type of medication and length of treatment. The side effects worsen with higher doses but tend to be less severe with time as the body adjusts to the medication. Antiepileptic drugs are usually started at lower doses and increased gradually to make this adjustment easier.

There are three types of side effects:

1. **Common or predictable side effects:**

These are generic, nonspecific, and dose-related side effects. They may occur with any anti-epileptic drug because it affects the central nervous system. These side effects include blurry or double vision, fatigue, sleepiness, unsteadiness, as well as stomach upset.

Idiosyncratic side effects:

They are rare and unpredictable allergic reactions that are not dose-related. Most often, these side effects are skin rashes, low blood cell counts, and liver problems.

Unique side effects:

Those that are not shared by other drugs in the same class. For example, phenytoin can cause the gums to swell and valproic acid can cause hair loss, weight gain, and tremor.

Unique side effects should be discussed with the patient before selecting the drug.

How medication is prescribed

How and when the drug is used may be an important factor in determining which drug patients prefer. If a patient's schedule makes it difficult to take medications several times a day, drugs given once daily may be preferred. The flexibility with which the drug can be prescribed is also important. For example, women with seizures related to menstruation may benefit from extra medication during their periods. Patients with nighttime seizures may benefit from extra doses at bedtime.

Follow-up visits

It may take several months before the best drug and dosage is determined for you. During this adjustment period, you will carefully be monitored through frequent blood tests to measure your response to the medication. It is very important to keep your follow-up appointments with your doctor and the laboratory to minimize your risk for serious side effects and prevent complications.

When seizures continue despite treatment for epilepsy, it may be because the episodes thought to be seizures are nonepileptic. In such cases, you should get a second opinion from a specialist and have EEG-video monitoring so the diagnosis can be reevaluated. In specialized centers, about 15% to 20% of patients referred for persistent, refractory, or intractable seizures ultimately prove to have nonepileptic conditions instead.

How long treatment lasts:

In some types of epilepsy, patients can be taken off treatment after a few years, while other types of epilepsy require life-long treatment. With few exceptions, patients who are seizure-free for a certain period should be reevaluated to determine whether the drug could be discontinued. How long the seizure-free period should be varies among the types of epilepsy and is controversial even for a given type. The decision to discontinue a medication also depends on more than the length of the seizure-free period. These factors include if the patient has normal neurological examination and if the EEG and brain scan (MRI) are/have been normal or not. If a medication is going to be discontinued, it should be weaned gradually to avoid triggering a seizure.

It may take time and careful, controlled adjustments by you and your doctor to find the combination, schedule, and dosing of medication to best manage your epilepsy. The goal is to prevent seizures while causing as few unwanted side effects as possible. Once the most effective medication program is determined, it is important that you follow it exactly as prescribed. Using a single antiepileptic medication is often better than using more than one medication. Single medication use causes fewer side effects and does not carry the risk of interacting with other medications. The chances of missing a dose or taking it at the wrong time are also lower with just one medication.

When treatment with one medication does not work, adding a second medication can sometimes improve seizure control. If you have several types of seizures, you may need to take more than one medication.

Phenobarbital

Phenobarbital requires one dose each day in most people. Infants may need two doses. Phenobarbital is available in liquid, capsule, and tablet forms. Primidone (Mysoline) is a drug that the body converts into Phenobarbital. It usually requires three doses per day. Phenobarbital is a barbiturate, which acts as a sedative or depressant. Phenobarbital is the antiepileptic usually used for newborns that have epilepsy. Partial seizures and generalized tonic-clonic seizures may be treated with Phenobarbital. Phenobarbital is successful in preventing partial and tonic-clonic seizures.

The Side Effects

Phenobarbital often makes children and older people:

- ✓ Hyperactive
- ✓ Impatient
- ✓ Unable to sleep
- ✓ Hostile

Phenobarbital has the opposite effect on young and middle-aged adults, who may feel:

- ✓ Lethargic
- ✓ Depressed
- ✓ Sleepy

Phenobarbital can alter your mood, behavior, thought processes, and ability to learn or remember things. These effects may be worse in older people. It may take time and careful, controlled adjustments by you and your doctor to find the combination, schedule, and dosing of medication to best manage your epilepsy. The goal is to prevent seizures while causing as few unwanted side effects as possible. Once the most effective medication program is determined, it is important that you follow it exactly as prescribed.

Effects

- ✓ Phenobarbitals affect your thinking and state of mind is a serious drawback to using the drug. Teenagers and adults taking the drug may feel depressed or irritable. It can cause memory loss and decrease your ability to learn. Children and older adults may feel restless and have trouble sleeping. Primidone is usually less effective and has more side effects than Phenobarbital (including depression and impotence).

Risk of birth defects

- ✓ Use of Phenobarbital during pregnancy increases the risk of minor birth defects in the newborn.

Pros:

- ✓ Phenobarbital only has to be taken once a day, making it a good choice if you have a busy schedule or have trouble remembering to take your drugs. Because Phenobarbital works very slowly, it may take weeks before Phenobarbital levels reach the proper level. However, missing a dose of Phenobarbital now and then usually does not affect the drug levels in your bloodstream. This is not true of other antiepileptic drugs.

- ✓ Phenobarbital is the cheapest of the broad-spectrum drugs used to treat a variety of seizures.

- ✓ Phenobarbital may reduce the effectiveness of birth control pills.

Ketogenic Diet

When the body burns fat, it creates substances called ketones. The ketogenic diet tries to force the body to use more fat for energy instead of sugar (glucose) by increasing fat and restricting carbohydrates. It is not yet clear how or why the ketogenic diet prevents or reduces seizures, but it has been shown to be effective in reducing epilepsy seizures in some children.

The ketogenic diet provides 4 grams of fat for every 1 gram of protein and carbohydrate. People on a ketogenic diet have to eat mostly fatty foods, such as butter, cream, and peanut butter. Foods such as bread, pasta, fruits, and vegetables have to be severely limited, and the person's total calories and fluids are restricted. At every meal, the food has to be measured carefully so that the right amounts of each food are given; even a slight departure from the diet can ruin its effect.

A person usually has to fast the day and night before starting the diet. The diet is gradually introduced over several days, so that the body can get used to the dramatic change. The person may feel tired and lack energy during the first few days.

Children are usually admitted to a hospital or epilepsy center when starting the diet so that they can be monitored. The ketogenic diet should always be under the supervision of a doctor and a dietitian. The ketogenic diet may sometimes be used to treat children with severe seizures who have not responded to antiepileptic medications. It has been especially effective in treating seizures related to Lennox-Gastaut Syndrome. It may be a good choice for a child when other treatments have failed to control seizures.

Doctors are not sure why the ketogenic diet helps prevent seizures, but the diet has prevented epilepsy seizures in many children who did not respond to treatment with antiepileptic medications, including children with Lennox-Gastaut syndrome. Another benefit of the ketogenic diet for some children is that it may reduce or end the need for medication and thus avoid the side effects that result from medication.

A recent review of the results from numerous studies of the ketogenic diet found that over half of children with seizures that do not respond to medications who follow the ketogenic diet have a 50% or greater decrease in seizures.[1] Some children have an even greater reduction. No one knows why the ketogenic diet prevents seizures in some children and not in others, or why it has been more successful with children than with adults. If you have a child with uncontrolled epilepsy, you may wish to discuss with your child's doctor whether a ketogenic diet might be an option.

The ketogenic diet may cause side effects in some people. These are not usually serious but may include:

- ✓ Dehydration
- ✓ Constipation
- ✓ Vomiting
- ✓ High cholesterol level
- ✓ Kidney stones
- ✓ Behavior changes
- ✓ Slower growth rates in children

People on the ketogenic diet may develop vitamin and mineral deficiencies unless they take vitamin and mineral supplements. The diet does not supply adequate amounts of the vitamins and minerals the body needs. Everyone on the diet needs close supervision by a doctor and a dietitian.

Despite the success of the ketogenic diet in some children, many doctors remain skeptical of its use. It may pose other health risks to your child, and it is extremely hard to follow. Until more found out about how the ketogenic diet works and what its effects are, some doctors may not recommend using it. If you are thinking about having your child try the ketogenic diet, keep in mind that it has several setbacks:

1. For the diet to prevent seizures, your child has to follow it exactly. The amounts and types of foods eaten have to be measured accurately, and preparing meals can take a lot of time.

2. The diet does not work for some children, no matter how closely they follow it.

3. The ketogenic diet is not a healthy eating plan for children or adults.

4. It does not taste very good.

5. People on the diet usually need to take vitamin and mineral supplements.

Felbatol (felbamate)

Felbatol (felbamate) can control certain epileptic seizures in adults and a rare form of epilepsy in children It is not known exactly how felbamate prevents seizures.

Felbamate may be used to prevent partial seizures in adults and children, although it has some very serious side effects. The medication used with caution in younger children since they may not be able to communicate symptoms of potentially serious side effects, such as chills or stomach pain.

Felbamate controls partial seizures in adults very well. In children with Lennox-Gastaut syndrome, felbamate may reduce seizure frequency, but not make the child seizure-free. However, it has made some children more alert and improved their quality of life.

The Side Effects

Common side effects of felbamate include:

- ✓ Nausea
- ✓ Vomiting
- ✓ Indigestion
- ✓ Headache
- ✓ Weight loss
- ✓ Sleep problems

Most of these problems tend to go away once the body adjusts to the drug. Weight loss and insomnia are common long-term problems; however, Felbamate should be used with extreme caution because it carries a significant risk of liver and bone marrow failure, which can be fatal. You or your child may need to be monitored frequently for signs of liver disease while taking the drug. A serious blood problem called aplastic anemia can also result from the use of felbamate.

Watch for early signs of liver, bone marrow, or blood problems, such as easy bruising, a change in skin color, prolonged bleeding, fatigue, fever, change in stool color, or a change in the color of the whites of the eyes. Despite these risks, felbamate may be used in some people because the drug has been successful in treating seizures that do not respond to other drugs (refractory seizures).

This is especially true of children with Lennox-Gastaut syndrome, which does not usually respond well to other drugs. However, because of its potentially life-threatening side effects, felbamate should be used only in those people for whom the risks of having seizures are greater than the risks caused by taking felbamate. Use of felbamate may make birth control pills less effective. If you are taking felbamate and birth control pills, you may be more likely to become pregnant. Felbamate is a drug that came out on the market recently. For some people, it may produce side effects or carry risks, including an increased risk of birth defects, which are not fully known yet. Report any unexpected side effects or problems to your doctor.

Lamictal

Partial seizures begin with abnormal electrical activity in a particular location in the brain, the effect of which depends on the area of the brain involved. The way that Lamictal works is unknown. It is believed that this drug alters the chemical impulses in the brain that cause seizures. Serious rashes requiring hospitalization and discontinuation of treatment have been reported with use of Lamictal. Approval of Lamictal based on a clinical study, published in the journal *Neurology*, demonstrating the effectiveness of the drug as an additional therapy in children who were still having frequent partial seizures (at least four per month) despite use of other epilepsy drugs.

Topiramate

Topiramate used several times by the patient each day. Topiramate comes in tablets and capsules, which can be opened and sprinkled on food. It is not known exactly how Topiramate prevents seizures.

The U.S. Food and Drug Administration (FDA) have approved Topiramate for use in combination with other antiepileptic drugs to control partial seizures in children ages 2 to 12. Topiramate is very effective in reducing partial seizures in most people who take it. It may also help control seizures caused by Lennox-Gastaut syndrome.

Side Effects

Common side effects of Topiramate include:

- ✓ Drowsiness
- ✓ Dizziness
- ✓ Fatigue
- ✓ Lack of coordination
- ✓ Loss of appetite
- ✓ Inability to concentrate or speak clearly
- ✓ Nervousness

Topiramate

Recently, Topiramate has been linked in some people to a potentially life-threatening condition called metabolic acidosis. Symptoms of metabolic acidosis include fatigue, lack of appetite, and rapid breathing (hyperventilation). If left untreated, metabolic acidosis can lead to death.

Topiramate is a new drug. For some people, it may produce side effects or carry risks, including an increased risk of birth defects, which are not fully known yet. Report any unexpected side effects or problems to your doctor. It may take time and careful, controlled adjustments by you and your doctor to find the combination, schedule, and dosing of medication to best manage your epilepsy. The goal is to prevent seizures while causing as few unwanted side effects as possible. Once the most effective medication program is determined, it is important that you follow it exactly as prescribed.

Tiagabine

Tiagabine increases the brain levels of a chemical messenger (neurotransmitter) called gamma-amino butyric acid (GABA), which may prevent abnormal electrical activity in brain cells.

Tiagabine is used in combination with other antiepileptic medications in adults and children over a Tiagabine age 12 to control partial seizures. When added to treatment with another antiepileptic drug, Tiagabine is sometimes effective in reducing partial seizures in adults and children over age 12 that have had trouble controlling seizures with other drug treatment. It is not helpful in reducing other types of seizures, such as primary generalized seizures or those in children with Lennox-Gastaut syndrome.

Side Effects

Common side effects of Tiagabine include:

Side effects - While, Topiramate is tolerated well by most people and has few serious side effects, a small number of people taking the drug may develop kidney stones. A recent warning links Topiramate use to metabolic acidosis in some people.

- ✓ Like most new drugs, Topiramate is expensive.
- ✓ Topiramate may make birth control pills less effective. A woman taking Topiramate may need to use another method of birth control to reduce her chances of becoming pregnant.

- ✓ Dizziness
- ✓ Headache
- ✓ Sleepiness
- ✓ Inability to concentrate
- ✓ Tremor

Vagus Nerve Stimulator (VNS)

Similar to a pacemaker, a vagus nerve stimulator (VNS) is a small device implanted under the skin near your collarbone. A wire (lead) under the skin connects the device to the vagus nerve in your neck. The doctor programs the device to produce weak electrical signals that travel along the vagus nerve to your brain at regular intervals. These signals help prevent the electrical bursts in the brain that cause seizures.

Once implanted in your body, the battery-powered device which has the ability to be programmed from outside your body by your doctor. You can also use a handheld magnet to turn the device on if you feel a seizure about to start. It takes approximately 2 hours to surgically place the VNS device in the chest. The vagus nerve stimulator can begin working right after the surgery (as soon as the doctor programs it). You may notice a slight bulge in the area under your collarbone where the device is, and the surgery will leave small scars on the side of your neck where the wire lead was placed and on your chest where the device was implanted.

Vagus nerve stimulation has been approved for use in treating people over age 12 with partial seizures who have not responded well to antiepileptic medications and are not candidates for epilepsy surgery VNS is used in combination with medication or surgery. While it does not eliminate the need for medication, it can help reduce the risk of complications from severe or repeated seizures. The vagus nerve stimulator reduces the frequency of partial seizures that do not respond well to medication and may make them less severe. It is used along with antiepileptic medications or epilepsy surgery to control partial seizures.

It appears from initial research that the benefits of VNS increase over time. After 3 months, 34% of people reported better control of seizures. After 12 months of VNS, 45% of people had fewer seizures—with 20% of those people reducing their seizure frequency by 75%.

For people who can sense when they are about to have a seizure, turning on the VNS can sometimes prevent the seizure. It may also shorten a seizure already in progress. Although the device has not yet been approved for use in children, initial studies show that it may be as effective in children as in adults. The children also showed improved alertness, mood, and memory; better school performance; and improved verbal communication skills.[3]

The vagus nerve stimulator is considered safe. Mild side effects occur in some people when the device stimulates the nerve. **The most common side effects include:**

- ✓ Coughing
- ✓ Throat pain
- ✓ Hoarseness or slight voice changes
- ✓ Shortness of breath

In children, vagus nerve stimulation may cause increased hyperactivity.

Vagus nerve stimulation is not a cure for epilepsy, and it does not work for everyone. It does not replace the need for antiepileptic drugs. Doctors are not exactly sure how or why the vagus nerve stimulator prevents seizures, and its long-term effects, if any, have not been studied. The U.S. Food and Drug Administration (FDA) have approved the vagus nerve stimulator, and this type of treatment is an area of ongoing research. It is becoming an accepted part of treatment for some types of epilepsy. Vagus nerve stimulation has not yet been approved for use in children under age 12, but early studies suggest that it may significantly benefit children with difficult-to-treat forms of epilepsy.

Keppra

There has been a dramatic reduction in seizures with Keppra. The study examined the effectiveness of the drug in 73 patients with an average age of 10 years who had epilepsy that had begun in childhood. At the start of the study, the participants averaged more than nine seizures a week. Following three months of treatment with Keppra in addition to their current therapy, that number dropped by 52%, to an average of 4.5 seizures a week. After about 16 months of follow-up, the average seizure frequency in patients treated with Keppra was reduced by 88%. Side effects included drowsiness or dizziness, and two patients reported aggressive behavior. Researchers say the drug was well tolerated and effective in about 90% of the patients studied. They also found that use of Keppra allowed the patients to reduce the total number of anti-epileptic drugs they took to manage their symptoms from an average of about three medications at the start of the study to two after the 16-month follow-up period.

Trileptol

In clinical studies, TRILEPTAL has been proven to help control partial seizures both by itself (monotherapy) and in combination with other anti-seizure medications (adjunctive therapy). During these studies, many patients taking TRILEPTAL experienced significantly fewer seizures.

Whether you have just been diagnosed or are currently taking other seizure medications, TRILEPTAL may help reduce the frequency of your partial seizures. Just look at some facts from clinical studies:

- ✓ 42% of patients who switched to TRILEPTAL saw a 50% or greater reduction in the frequency of partial seizures

- ✓ 91% of newly diagnosed adults taking TRILEPTAL experienced a greater than or equal to 50% reduction in frequency of partial seizures
- ✓ 89% of newly diagnosed children experienced a greater than or equal to 50% reduction in partial seizure frequency after taking TRILEPTAL

Favorable Safety and Tolerability Profile

What may surprise you about TRILEPTAL (Oxcarbazepine) is that it is effective and has a favorable safety and tolerability profile. It is not associated with the bothersome side effects of other antiepileptic medications like:

- ✓ Unusual hair growth
- ✓ Weight gain
- ✓ Coarsening of features

If you have tried other antiepileptic drugs, you may have needed to undergo regular blood tests to check for liver enzyme abnormalities. That type of monitoring is not required with Trileptol.

Easy to take tablet or liquid, used twice a day. Trileptal is also convenient to use. To begin with, you have your choice of liquid or tablet. Whichever you choose, you take it just two times a day. It can be taken with or without food. Only a Physician can decide if Trileptal is right for you or your child. Find out if Trileptal can help increase control of your seizures so you can enjoy more of life's pleasures every day.

Important Safety Information

Trileptal (Oxcarbazepine) is indicated for use as monotherapy (by itself) or adjunctive therapy (taken with other medications) in the treatment of partial seizures in adults and children 4 years of age and older with epilepsy.

The most common side effects (occurring in at least 5% of patients taking Trileptal) were dizziness, sleepiness, double vision, fatigue, nausea, vomiting, in coordination, abnormal vision, abdominal pain, tremor, indigestion, and abnormal gait--these were typically mild to moderate in severity. Alcohol consumption may increase sleepiness. You should avoid driving or operating machinery until you determine how Trileptal affects these activities.

A condition called hyponatremia (low blood sodium levels) has been observed in some patients treated with Trileptal. If your doctor thinks you may be at risk of hyponatremia, he or she may choose to monitor your sodium blood levels. You should inform your doctor if you have used the epilepsy drug Carbamazepine in the past. Twenty-five percent to 30% of patients who have had allergic reactions to Carbamazepine may experience allergic reactions to Trileptal

Use of Trileptal with birth control pills may render birth control pills less effective. It could make the effect of birth control pills as effective as taking sugar pills. Additional forms of contraceptives are recommended when using Trileptal.

The recording of brain waves by electroencephalography is an important diagnostic test for epilepsy. Other diagnostic technologies include CAT scan and magnetic resonance imaging (MRI). Standard treatment of epilepsy is with anticonvulsive drugs, such as Carbamazepine, Phenytoin, and

Valproate; it requires a careful analysis of seizure motor activity, anatomical cause, precipitating factors, age of onset of the disorder, severity, daily rhythms, and prognosis.
Some cases of childhood epilepsy (which is often eventually outgrown) have been successfully treated with surgery or a very high-fat "ketogenic" diet. The diet results in a natural buildup of ketones in the body, which appear to inhibit the seizures. First aid, such as cushioning the head, used to prevent the person from self-inflicted injuries during seizures. With proper medication, most people with epilepsy live normal lives. Repeated seizures that lead to unconsciousness, however, appear to be associated with damage to the hippocampus in the brain and sudden unexpected death.

To help temporal lobe epilepsy a procedure called, "What Is a Temporal Lobe Resection can be performed on the patient."

A temporal lobe resection is an operation performed on the brain to control seizures. In this procedure, brain tissue in the temporal lobe is resected, or cut away, to remove the seizure focus. The anterior (front) and mesial (deep middle) portions of the temporal lobe are the area's most often involved.

Temporal lobe resection may be an option for people with epilepsy whose seizures are disabling and/or not controlled by medication, or when the side effects of medication are severe and significantly affect the person's quality of life. In addition, it must be possible to remove the brain tissue that contains the seizure focus without causing damage to areas of the brain responsible for vital functions, such as movement, sensation, language, and memory. Candidates for temporal lobe resection undergo an extensive pre-surgery evaluation- including seizure monitoring, electroencephalography (EEG), magnetic resonance imaging (MRI) and positron emission tomgraphy (PET). These tests help to pinpoint the seizure focus within the temporal lobe and to determine if surgery is possible.

A temporal lobe resection requires exposing an area of the brain using a procedure called a craniotomy. (*Crani* refers to the skull and *otomy* means, "to cut into.) After the patient is put to sleep, the surgeon makes an incision (cut) in the scalp, removes a piece of bone, and pulls back a section of the dura, the tough membrane that covers the brain. This creates a "window" in which the surgeon inserts special instruments for removing the brain tissue. Surgical microscopes also are used to give the surgeon a magnified view of the area of the brain involved. The surgeon utilizes information gathered during the pre-operative evaluation-as well as during surgery-to define, or map out, the route to the correct area of the temporal lobe.

In some cases, a portion of the surgery is performed while the patient is awake, using medication to keep the person relaxed and pain-free. This is done so that the patient can help the surgeon find and avoid areas of the brain responsible for vital functions. While the patient is awake, the doctor uses special probes to stimulate different areas of the brain. At the same time, the patient is asked to count, identify pictures or perform other tasks. The surgeon can then determine the area of the brain associated with each task.

After the brain tissue is removed, the dura and bone are fixed back into place, and the scalp is closed using stitches or staples. The patient generally stays in the hospital for 2 to 4 days. Most people having temporal lobe resection surgery will be able to return to their normal activities, including work or school, in 6 to 8 weeks after surgery. The hair over the incision will grow back and hide the surgical scar. Most patients will need to continue taking anti-seizure medication for two or more years after surgery. Once seizure control is established, medications may be reduced or eliminated. Temporal lobe resection is successful in eliminating or significantly reducing seizures in 70% to 90% of patients.

The following symptoms may occur after surgery, although they generally go away on their own:

- ✓ Scalp numbness
- ✓ Nausea
- ✓ Feeling tired
- ✓ Depressed
- ✓ Headaches
- ✓ Difficulty speaking, remembering or finding words
- ✓ Continued auras (feelings that signal the start of a seizure)

The complication rate with temporal lobe resection is low, but there are some risks, including:

- ✓ Risks associated with surgery, including infection, bleeding and allergic reaction to anesthesia
- ✓ Failure to relieve seizures
- ✓ Changes in personality or mental abilities
- ✓ Pain

Chapter 7: How to Boost Your Self-Esteem

Learn How to Boost Your Self-Esteem

Self-esteem is linked to how good you feel about yourself and how well you value yourself. Building self-esteem is a first step towards happiness and a better life. Self-esteem increases your self-confidence. If you have confidence, you will love yourself and begin to respect yourself. If you respect yourself you can respect others, improve your relationships, your achievements and your happiness....

Building self-esteem is a first step towards a happy, healthy, and productive life. Yet having low self-esteem is perhaps the most common flaw of our humanity. Having self-, esteem will help build your confidence. If you have self-confidence, you will feel a self-worth and respect yourself as a person. If you respect yourself you can respect others, improve your life by improving your relationships with your love ones, friends and co-workers. You will be able to achieve your goals in life and obtain true happiness in your life.

It is not easy to change or improve. As with most efforts of change, you get out only as much as you put in. Determination is very important. It will take time; you may be able to see quick improvements by making small changes in your life or thinking. To make permanent changes you have to work hard and continue in your hard work. Building self-esteem is one thing but keeping a high level of esteem requires that you use the things you learn in this book into your life and make them a part of your life. Low self-esteem can cause people to develop depression, unhappiness, insecurity and a poor confidence level. When you have low self-esteem ever little mistake that occurs in your life is taken to heart. When people criticize you, when you make mistakes, a joke directed toward you, it would cause you to run away from every opportunity that comes your way, and every challenge will seem impossible to accomplish. This will cause you to feel stagnant and lose meaning and direction in your life.

You can develop high self-esteem, just like learning to read or dance. Secondly, people do not understand the importance of having high self-esteem. I cannot even begin to stress the importance of having high self-esteem; it is the key to having mental, physical and spiritual strength.

The first stage of developing strength is learning to love yourself and your life. You need to learn to be grateful of what God has given you. You need to let go of all those angry emotions inside. Holding anger inside yourself will not help you, it will only hurt you. The past is the past; you can only change the present. You need to love yourself by accepting all your faults and putting the past behind you, but if, you focus on your faults than you will only experience an unhappy life. You need to think positively and focus on your accomplishments.

To live with a happy state of mind, you need to have high self-esteem. You need to feel that you are no different from anyone else and that you can be the person you set in your mind to be. You need to reconstruct your life. You need to put yourself in a lifestyle that will make you happy and bring you as little stress as possible.

To begin the healing process you need to develop strength, wisdom, confidence and knowledge. If you can develop these qualities, you will achieve all your goals and dreams. First, you must focus on the goals and dreams you want to fulfill. I am going to teach you the true meanings of having strength, wisdom, confidence and knowledge. I will help shoe you how to obtain and use them. These four steps will help you live a happy life and gain high self-esteem! Below are four steps to high self-esteem.

1. **Strength**-the development of strength in the inner body begins in the mind. The inner body is our mind, soul and spirit. How we think and program our minds to work, helps us build mental, physical and spiritual strength. Our strength comes from how we feel about ourselves. The higher our self-esteem, the stronger we feel and in turn, we can do more for ourselves. Yet, if you have the strength and motivation, you can make the present anything you want. To free all your negative emotions that are holding you back, you have to say to yourself. I accept myself for who I am and that I am unable to change the past. Nevertheless, I can change my future because I love myself and refuse to hurt myself by drowning in my own self-pity. You cannot rely on others. You need to learn to rely on yourself. You have to believe in yourself, develop a sense of pride in yourself. It does not matter what others think about you, what matters is how you think about yourself. God put us on this earth to love others, not to hurt

ourselves and take our anger out on others, who are usually the people we care about the most and ourselves.

2. **Knowledge** - is the second part of the process of change. In order to higher your self-esteem and grow as a person you must be open minded to suggestions others may give. We may not always agree with other people's suggestions, yet it is always wise to listen to what others have to say. Some individuals may try to be controlling and may get frustrated if we do not act on what they have to say. You should to set these people straight and tell them; I will listen to what you have to say; however, that does not necessarily mean I am going to agree with you. I have my own mind, too and I need to do what is best for me. We learn from each other and we acquire knowledge from the world around us that we should pass along to others by helping them. We need to take our experience and use it in our present life now, including the mistakes we

have made in life. The mistakes we have made are where we get most of our knowledge that helps us become stronger individuals. What weakens us when we repeatedly make the same mistakes? Do not pity yourself for the mistakes you made in life or imperfections. Studies have shown that people who have negative attitudes are more like to live chaotic lives. They are more likely to become mentally or physically ill with extremely debilitating or life threatening illnesses. Many people have a hard time focusing on the positive because they allow their negative sides to consume them... I firmly believe that focusing on the negatives will cause seizures. Say to yourself, OK, what I have learned from these mistakes or from my shortcomings. Taking what you have learned and using it to help others is the best therapy. When you help, you feel a sense of accomplishment and self-worth. You are overlooking any negative characteristics because you are too busy focusing on helping others'.

3. **Confidence** - our confidence comes from our self-esteem. To have high self-esteem we need to feel good about ourselves, to get to this point in life you need to begin by starting to do things in life to make yourself happy by focusing on the future, creating direction in your life. Begin by planning short and long-term goals for yourself and confidence level will rise. It worked for me. When I started accomplishing some of my short-term goals, I had more self-respect. I developed a greater sense of pride and my inner strength and self-worth increased.

4. Wisdom - comes from your sixth sense. We all have five senses, our sight, hearing, smell, taste, touch, yet I believe wisdom to be our sixth sense. Wisdom understands the inner signals and the directions that your body sends out to you, becoming aware of what your body is trying to tell you. Your sixth sense always leads you to the right answers. It is up to us to learn to understand our inner self (spirit) and to follow the signals it sends out to us. Listening to what our inner self has to say is essential. For example, have you ever felt like you had a feeling something was the right thing to do. You need to learn to understand your mind, so you can understand your inner soul and all the wonderful things it is capable of doing. When we listen and act on the signals our body, it gives us a stronger understanding to our body as a whole. Spiritually you can give your body what it needs. We feed our body food to survive on a daily basis. Spiritually we need to feed our body with love, understanding and different forms of relaxation, such as meditation. I strongly suggest to everyone that you start with at least five minutes each day with some type of relaxation exercise. Either in the morning when you start your day, the afternoon if you are able to or at night before bed to release the tension that has built up throughout the day. Each week you should add five minutes until you get to hour each day.

When you do these things, you increase your level of strength, wisdom, knowledge and confidence. By having a high level of strength you feel as though, you can conquer the world. This helps you decrease your stress level.

Once you accept yourself, you can cope with the world around you and accept the fact that you can do everything you expected to do in life. Nevertheless, to accept that yourself you first have to love who you are and be proud of the person you have become. There are many things in life you are capable of doing, but you must develop the motivation and the will to get out there and JUST DO THEM!

Things to Remember:

1. *Listen to your inner voice and follow your heart. Your heart always knows what is best for you.*

2. *Focus on the positive, believe in yourself, strive for the best, and expect the best-you ought to*

have nothing less. Just because you have epilepsy, it does not mean that you do not deserve the best.

3. *Let to give to others. The satisfaction of helping others hell boost your self-esteem*

4. *You cannot change the fact that you have epilepsy. Accept your disorder and focus on your future encouraging yourself to better yourself and achieve great things for the future.*

5. *Give your best effort and reward yourself for your efforts.*

EPILEPSY

The Most Important Secrets
You Must Learn
In Order To Live, Learn, and Be Happy With Epilepsy

Chapter 8: Getting On with Your Life

Living with a Disability, Everyday of your Life is Tough.

Living with a disability, every day of your life is tough. For many who have a disability they tend to hold their emotions inside. The emotions that develop tend to build up inside them until they are unable to deal with their emotions any longer. When you ignore your emotions and hold your emotions inside you set yourself up where you can easily fall into depression. This can happen when one focuses on the negative aspects of their disability and by pitying themselves.

Living with a disability can be difficult if you do not accept the disability into your life. When you accept the disability into your life, you must first realize that there is no such thing as a perfect person. We triumph each day of our life trying to master how to solve the daily troubles that come our way, and how to overcome the problems that have already occurred in our lives. You need to grasp the notion that no one on this earth is perfect and there is no need to feel a sense of embarrassment because you have a disability. If you look into any person's closet, you will find plenty of deep dark secrets. Overlooking your problems and not dealing with them is the easy way out, yet to face that dilemma and deal with it is an accomplishment.

Accepting our problems and dealing with them helps us grow mentally, physically and spiritually. One should not feel ashamed because they have a disability. When I opened up, telling people about my disability, I was shocked to find out how many people had some of disability or knew someone who had a disability. Many individuals are uneducated about the different disabilities that effect individuals in our society and look at the people who have the disabilities as weird or think of them as different. Nevertheless, both statements are untrue. I believe God puts obstacles in our lives to strengthen us. When we are young, we have people in our lives that help to mold us. They help us develop the strength, wisdom, and knowledge we need to survive in this world. Yet if we become dependent on these people, we cannot survive and live the productive life that God has given us on this earth to enjoy. You must realize that they put everyone on this earth here for a reason. We need to pass on what we have learned along to others.

I believe it is just selfish and pure laziness when we pity ourselves because we have a disability. You need to take your problems and learn how to cope with them so you can help other people. You are not alone, if you feel you are unable to straighten out your life the way you want it to be. There is no reason why you should not live a happy and healthy life just because you have a disability.

You need to accept your disability into your life and look at it positively. To do this you need to open your heart and feel what you emotions are trying to tell you. Your heart will never lie to you because the heart only holds the truth. You need to develop courage so you can ask your heart, why you refuse to accept the fact that you are disabled. Usually when we chose to hide things about ourselves, it is because we are embarrassed of whatever we are trying to hide. You should not be ashamed of having a disability. People with conditions, diseases, and disorders are constantly coming out into the open. They are learning to talk about the problems in their lives. At the same time, these people are educating society about their problems and healing the scars that lye inside them.

Society is becoming less fearful toward all the conditions that unfortunate individuals have to live with each day of their lives. Society is now trying to help through sponsoring support groups, research studies, and so forth. Disabled individuals need to learn to accept what they have and learn to do something about it. Nothing is going to get better until you learn to help yourself and help others.

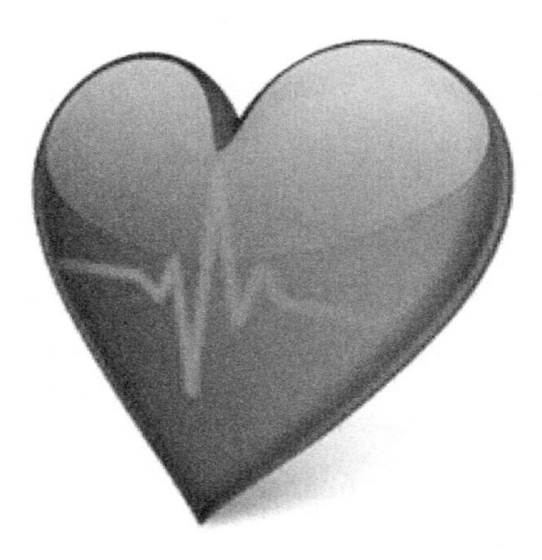

Chapter 9: Learning to Love Yourself Again

Self-love may be the greatest and most important love you will ever experience in your life. Learning to love yourself may seem like a difficult task to achieve. Many people try to show a facade. They go around acting happy like nothing is wrong and life could not be better, but inside hey hurt emotionally. They have emotional chains wrapped around their hearts. Love cannot enter their hearts. Their hearts are locked so they cannot let the gift of love flow from their hearts into someone else's heart. They walk around with pain and emotionally suffering because they are angry, perhaps because they have epilepsy. They may feel different because others can drive and they cannot. For whatever the reason, they are drowning in hidden pain that no one knows about, but them.

Now you can stop all this. It is time to find the key and unlock the locks wrapped around your heart. It is time to learn to love yourself once again. An important way to love yourself is to take care of yourself emotionally, physically and spiritually. You need to take time out for yourself. Go out and get a massage, a facial, a pedicure, or something that strikes your interest that you enjoy to do.

You need to go out and have fun. Go out to dinner, go dancing, and go see a movie go to a baseball game, take a walk on the beach. Set aside time and plan a vacation. Go vacation somewhere relaxing and fun. Do not analyze yourself. Do not criticize yourself. You may get angry and call yourself stupid or say I cannot do anything right. You need to stop those negative thoughts and replace them with positive thoughts. For example, "I made a mistake that is okay no one's perfect." We learn by our mistakes. When you begin to think negative, immediately stop yourself, and change your thought into a positive one. If you are telling yourself that you are a failure, you are much more likely to fail than if you picture yourself as a success. There is nothing wrong to give yourself a pat on the back. You need to give yourself a little boost to be a winner.

Reward yourself when you do something to better yourself, give yourself something special. Take time to each day to tell yourself, "I accept that I have epilepsy", "I am no different than anyone else", "I am beautiful and bright" or "I love my myself," "I am loving, caring, and worthy of love" or "I believe in myself." Write these quotes down and put them where you can see them everyday

You must realize that what is most important is not what others think about you it is how you feel about yourself. In order to feel good about yourself, you need to accept yourself. You must look at yourself positively and realize that there is no such thing as a perfect person.

What is most important in our life is that we try to master the daily troubles that come our way, and overcome the problems that have already occurred in our lives. Now this should be considered a triumph. If you are unhappy with yourself then you need to do something about it. Ignoring your problems and emotions, not dealing with them, is the easy way out, to face them and deal with then are accomplishments. Accepting ourselves helps us grow mentally, physically and spiritually.

Remember, it is not how we look on the outside that is important. It is how we feel on the inside that matters the most. In this life in order to survive, it is important that we accept ourselves for who we are, feel good about ourselves, and carry an inner strength so we can live a healthy, happy, and productive life.

When we are young, we have people in our lives that help to mold us. They help us develop the strength, wisdom, and knowledge we need to survive in this world. Yet many of us forget what we learned and focus on what is unimportant, such as how we look on the outside.

The results are in the end you are going to end up feeling

emotionally drained and unable to live a productive life. Everyone is on this earth here for a reason. We need to pass on what we have learned along to others.

It is selfish and pure laziness to pity ourselves because you may not have that model figure. You must learn to accept and love yourself. You are not alone, many women feel unhappy with themselves.

To live with and accept yourself, you need to open your heart and listen to what your heart is telling you. Your heart will never lie to you because it only tells the truth. You must have the courage to ask your heart why you refuse to love yourself the way you are. Usually when we are unhappy with ourselves, it is because we are embarrassed about whatever we are trying to hide. Your body is not something you should be ashamed of having. You need to help yourself by accepting yourself and loving yourself for who you are. This is the first step in order to heal yourself and feel good about you. Ask yourself the question, "What level of self-approval you have I reached on a day-to-day basis?" I have listed seven steps to help you learn to love and accept yourself.

STEP 1- Accept yourself for who you are and learn to love yourself for whom you are a person

STEP 2- Understand yourself mentally, physically and spiritually

STEP 3- Learn to control your mind, body, and emotions

STEP 4- Strengthen your inner self and make it apparent to others

STEP 5- Begin changing what you do not like about yourself

STEP 6- Notice the change in your self-esteem and self-confidence

STEP 7- Have a tremendous amount of pride in yourself

People realize that to change and strengthen themselves they must accept themselves and learn how to live with themselves in a productive manner. You need to look at life in a positive way. You need to say to yourself, "OK, I'm not happy with the person, I have become. I need to change and this is what I am going to do about it." Stop being lackadaisical. This is the first step to healing and strengthening our souls and self-esteem. Be proud in who you are. Be thankful each morning that you can wake up and feel the warmth of the sun and the beauty that surrounds us all.

Love yourself, be good to yourself, and treat yourself good. The more you love yourself, the more you will be able to give love to others - and the more others will want to be around you and give back to you. Loving yourself will help the lives of others around you, as well as your own life.

Chapter 10: Say Goodbye to Stress and Say Hello to Happiness

Life can be a circus! It is very hard to avoid stress. Stress is something we all experience in our daily lives, and depending how you manage your stress can sometimes be overwhelming. If you let stress linger, it can be harmful and even be one of the causes of your seizures. It is important to recognize early signs of stress and do something about them. Doing so can drastically improve your quality of life.

One of the first skills I learned in managing stress is how to get rid of stress with healthy and productive techniques:

Every day we wake up to a new day with new experiences to explore and responsibilities to accomplish. Remember, Rome wasn't built in a day. If you do too many things at once, you will become overwhelmed. In the end, you may not accomplish anything because you are so stress you cannot think straight and in the end, you can cause yourself a seizure. Make a list of the things you have to do. Then work on them one at a time, checking them off, as they are finished. Do the most important ones first and take care of the rest later. The best strategy for keeping stress out of your life is learning how to relax. Take time to tune out worries and responsibilities.

Focus on relaxing and enjoying life.

These techniques will help you feel at ease no matter where you are or what you are in the middle of at that moment. Most stress relieving techniques can be done with hardly any or no instruction. Doing these, techniques on a regular basis will help you cope with life so you can enjoy yourself and live a healthy productive life.

As I was working to accept myself and love myself, I noticed myself changing emotionally. I felt better about myself. I would look in the mirror and be proud of the person I was seeing. Start waking up each day. This will give you some extra time to prepare for the day ahead. Spend 30 minutes just relaxing, do something that relaxes you. I like to get up make a cup of coffee and go in front of the computer and read my e-mail. That would help me relax before I started the day. You could take a walk, read something uplifting, and think about what you need to be done. After taking time for yourself in a peaceful setting, you will feel better equipped to handle whatever lies ahead

We need to understand how our bodies work and listen to its messages. When a stressful situation comes up, decide that you are not going to let it stress you. If every stressful situation that occurs upsets you, you will end the day feeling drained and unhappy. Save your energy for those situations that are meaningful.

Remember sleep is important. The outcome of a full night's sleep is amazing. You wake up feeling energized. When you are well rested, you have a more positive outlook on the day, including the things you must accomplish.

When I was young, I thought I could do everything. I realize now, as a young woman, there is no reason for me to lower my expectations, but there is also no reason to push myself over the limit. No one on this earth is a 100% perfect.

We all have our faults. I work all the time by trying to make myself into a stronger human being emotionally, physically and spiritually. Working on myself makes me feel like I can climb over the barriers that are put in front of me at times. Everybody has different characteristics that make up their personalities.

Epilepsy is just one part of me. I cannot change the fact that I am epileptic. I have to accept it the fact that I have epilepsy and learn to live with it productively. You will succeed and become a better person, if you think positively and productively about yourself. You will feel the strength in yourself to accept yourself and love yourself for who you are. Think about who you and where you are headed in life. It is up to you to make something of yourself.

Here are exercises to help you improve yourself physically and mentally. I call this the spirituality builder. It uses all your muscles and is designed to help you develop strength and flexibility in the body, increasing circulation, and stretching. These exercises will help you feel good and at peace with yourself. I do these exercises frequently. It helped me feel like I had full control with myself. We may not have control over ourselves when we have seizures; however, we do have the power to figure out how to deal with our epilepsy so it does not become a problem in our lives.

EXERCISE ONE

1. *Lie on your back with your arms at your sides. Adjust your body to a comfortable position.*

2. *Slowly relaxing all your muscles in your body starting with the feet and then working your way up to your head.*

3. *Keep your eyes closed, concentrating on what you see inside yourself. Focus on what you feel and what you want to feel like.*

4. *Slowly raise your body upright, bending forward*

5. *Then bend to your right side*

6. *Then to your left side*

7. *Then back to step one position*

8. *Remember to take deep breaths during this exercise and breathe slowly.*

9. *Do this for a minimum of fifteen minutes a day.*

EXERCISE TWO

This exercise should be done twice a day for 5-15 minutes in a quiet room free from disturbance.

1. *Rest on your back with head and neck comfortably supported*

2. *Rest hands on upper abdomen, close your eyes, and settle in a comfortable position.*

3. *Breathe slowly, deeply and rhythmically. Inhalation should be slow, unforced, and unhurried. Silently count to four, five, or six, whatever feels right for you.*

4. *When inhalation is complete, slowly inhale through the nose. Count this breathing out, as when breathing in. The exhalation should take as long as the inhalation. There should be no sense of strain. If initially, you feel you have breathed your fullest at a count of three, which is all right. Try gradually to slow down the rhythm until a slow count of five or six is possible, with a pause of two or three between in and out breathing.*

5. *This pattern of breathing should be repeated 15 or 20 times and since each cycle should take about 15 seconds; this exercise should take about 5 minutes to do.*

6. *Once the mechanics of this exercise have been mastered, try to introduce thoughts at different parts of the cycle. On inhalation try to sense a feeling of warmth and energy entering the body with the air. On*

exhalation sense a feeling of sinking and settling deeper into the surface, you are lying on.

On completion, do not get up immediately but rest for a minute or two, allowing the mind to become aware of any sensations or stillness, warmth, heaviness etc. Once mastered, this exercise can be used to help you cope with any situation, so you do not become over agitated.

EXERCISE THREE

Often tension is focused in the muscles of the body itself, and the following exercise itself can release such tightness and allow the mind to be at ease. It is best to begin this exercise with a few cycles of deep breathing.

1. *Lie down or sit down in a reclining chair.*

2. *Avoid distractions and wear clothes that are comfortable.*

3. *Starting with the feet, try to feel or sense*

that the muscles of the area are not actively tense.

4. *Then deliberately tighten the muscles, curling the toes under and holding the tension for 5 or 10 seconds.*

5. *Then tense them even more strongly for a few more seconds before letting go of all the tension and sensing the feeling of release.*

6. *Try to consciously understand what this feels like, especially in comparison with the tense state in which they were held.*

7. *Exercise the calf muscles in the same way. First sense the state the muscles are in, then tense them, hold them in position, and then tense them even more before letting go. It is a positive sense of relief. If cramping occurs, stop tensing that area immediately and go on to the next area.*

8. *After the calf go on to exercising the knees, then the upper leg, thighs, buttocks, back, abdomen, chest, shoulders, arms, hands, neck, head and face. The particular order is irrelevant, as long as these areas are exercised the same way.*

9. *Some areas may need extra attention. For example, in the abdomen the tensing of the muscle can be achieved either by contracting (pulling in the tummy) or by stretching (pushing outwards). This variation in tensing is suitable to many muscles in the body.*

10. *There are between 20 and 25 of these areas depending how you go about it. Give each about 5-10 seconds of tensing and another 5-10 seconds of letting go. It should take 8-10 minutes to complete this exercise. After the exercise, try to relax for a couple of minutes.*

11. *Focus the mind overall body. Sense it as heavy and content, free of tension. You can do this by doing a few cycles of deep breathing.*

Chapter 11: Our Dreams Are Our Future

Dreams are the pathways to our inner souls and come from our subconscious mind. While we are sleeping, our body tries to send messages about the wants and needs our body. A person's dreams can give a sense of direction in life. Even though we cannot remember, a majority of our dreams when we wake up still have an impact on the way we think and function. Sigmund Freud believed that dreams are expressions of unfulfilled wishes and desires.

Dreaming plays an important role in our lives. Studies have shown that people who are repeatedly awakened at the beginning of dream periods for several nights become irritable and have difficulty concentrating. If your bodies' natural sleep cycle has been interrupted and has been deprived of dream sleep, your body will compensate by providing proportionately more dream sleep at the next dream sleep opportunity. Research shows that a healthy sleep is needed for a person's body to restore itself. Some scientists believe that adequate dream sleep is equally important because it enables the brain to recharge.

Medical research has not proven this testimony. Usually, when a person is awake, their brain waves will show a regular rhythm. When a person first falls asleep, the brain waves become slower and less regular. They call this sleep state non-rapid eye movement (NREM) sleep. NREM sleep consists of stages. There are four stages and each stage is a progressively deeper stage. The deeper the sleep, the more your body restores itself. Stage one sleep is the transition from wakefulness to sleep. Restoration begins in stage two, but is more significant in stages three and four, sometimes called delta sleep.

After an hour and a half of NREM sleep, the brain waves begin to show a more active pattern again, though the individual is in a deep sleep. These sleep states are called rapid eye movement (REM) sleep, is when dreaming occurs. A person typically experiences a brief arousal from sleep and returns to stage two sleep after dreaming. This sleep cycle has begun again. The length of time in each of these stages differs throughout the night, with most REM sleep occurring during the later sleep cycle.

Dreaming and fantasizing give you a feeling of serenity and inner peace. Fantasizing has a positive impact on you and your body. When you fantasize, you put yourself in a state of consciousness that lies between reality and the world of dreams. The imagination roams freely, although usually guided by mostly unconscious urges, concerns and memories. Fantasies help us find out what type of ambition we have and the people we want to become in life; it takes us into another world where we can do and become anything we want. It allows us to relax and joyfully think about the various scenarios that would make us happy.

While growing up, I always enjoyed sitting and closing my eyes and fantasizing about something relaxing. It seemed when I understood my body and mind, I was able to make them function as one and then release all negative thoughts and feelings that were putting unnecessary stress on me. My stress level decreased and so did the total amount of epileptic seizures I was having monthly.

During this time of dreaming and fantasizing, you can focus on anything you want, including making goals for yourself on how you are going to live your whole life. It is important that you always dreamed about positive things. Do not think negatively or feel sorry for yourself. Look at yourself as a fighter and an achiever. Do not let anything get in your way.

Dreams can stay in your mind, no one has to know about them, and you can record them, in your daily diary, where you write down your significant dreams and fantasies. These dreams and fantasies can be used as motivators to help you work on loving yourself. Your dreams and fantasies can help you plan your life five or even ten years from now and you can use your dreams to strengthen your inner self. When life seems too stressful to handle; close your eyes and let your mind take you somewhere you can relax and fantasize.

We have all experienced dreams, the gateway into a chaotic territory of joy and embarrassment, excitement and fear. Dreams are the pathways to your inner soul, your soul that knows what your mind and body needs. Reach out and get in contact with your soul because it is necessity that you take the time out to understand what your mind and body crave.

Keeping a journal is a very successful tool for learning about yourself and finding out things that you never realized about yourself before. I write everything down on paper from my short-term goals to my long-term goals, to my dreams and fantasies. This helps increase my inner strength by keeping me in touch with who I am and what I need to do with my life. By writing, everything down expressing how you feel you can understand yourself better and the needs of your body mind and soul.

Now onto the four steps to strengthening yourself:

1. Start with writing a journal. Create a journal that will help you strengthen and understand yourself, and make you feel better about yourself as a person.

Use this format to set up your journal.

- ✓ MY DREAMS WHAT THIS DREAM MEANT TO ME
- ✓ HOW WILL I USE THIS DREAM TO MOTIVATE ME?

- ✓ MY FANTASY WHAT THIS FANTASY MEANT TO ME
- ✓ HOW WILL I USE THIS FANTASY TO MOTIVATE ME?

3. Create a list in your journal. Keep a list of the dreams and fantasies you have that mean something special to you. Write down how they can motivate you in this program and how they relate to helping yourself.

4. Do they relax you? Do dreams help you understand yourself and better yourself as a person? Do they give you hope for the future? While you write in the dream portion of your journal, answer the questions above. This is another way to strengthen yourself and understand your personal make up. These exercises will help strengthen your mind, body and soul in many ways

Chapter 12: Let Your Confidence Be Your Strength

Self-confidence-like being rich - it is something we think everyone else is but us. Yet feeling insecure about ourselves is perhaps the most common problems of humanity. The majority of people in society think, "You have to be born with it." This is not true. You can learn how to be confident, just like learning how to cook or drive a car.

Self-confidence all boils down to how we feel about ourselves. Our self-confidence develops and affected from the moment we are born, to how we are raised and things we experience during our lifetime. Many of us tend to judge how well we are doing in life by what society thinks it is right or what the people who center our life think.

We all have standards and expectations that we all tend to try to live up too. If we do not meet up to our standards, which most of us do not, our self-confidence begins to slowly decline. Many of us try to have unrealistic standards that we try to live up to and we tend to lose sight of the fact that we cannot be perfect.

We need to accept ourselves, love ourselves and be happy with who we are as people. We were valued for just being people, for just being in this world. We often believe we must continually try to prove to other people that we are worthy of them or on the same level as them. We often spend too much time doing that and we lose sight of the fact that we are fine just the way we are and that the only person that needs to be happy with them is themselves even though we are not perfect people.

In order to gain self-confidence you need to believe in yourself. If you believe in yourself then you can succeed in anything you put your mind to. You may not succeed the first time you try, but you have to keep trying until you do succeed. Quick success does not exist in our society. Achievements only come to those who strive hard to get them. You get nowhere in life if you do not push yourself. You need to create a lifestyle that is right for you and nobody else. Do not settle for anyone else's lifestyle or for a lifestyle that is beneath your standards.

My "10 step plan" To increasing confidence:

1. *Begin a journal. Ask yourself what is making you feel like you cannot get to the point in life you want*

to reach. Write any thoughts that come to mine. Also, write down what self-confidence means to you.

2. *Remember the past is over; you can only change the future.*
   Write down 10 positive things about yourself. Go through your journal and look at all the positive things about yourself. Concentrate on your strengths. These are the reasons you should love yourself and have high self-esteem and self-confidence in yourself. Give yourself credit for everything positive you have written about yourself. Remember, you are somebody special.

3. *Accept yourself and learn to love yourself for whom you are a person. Everyone has his or her own unique qualities and characteristics. We are all born differently for a reason. Do not compare yourself to others.*

4. *Understand yourself mentally, physically and spiritually. Take*

some time to relax by yourself in a quiet room... Rest on your back with head and neck comfortably supported.

5. *Rest hands on upper abdomen, close your eyes and settle in a comfortable position. Breathe slowly, deeply and rhythmically. Inhalation should be slow, unforced and unhurried. Silently count to four, five or six, whatever feels right for you. When inhalation is complete, slowly inhale through the nose. Count this breathing out, as when breathing in. The exhalation should take as long as the inhalation. There should be no sense of strain. If initially, you feel you have breathed your fullest at a count of three, which is all right. Try gradually to slow down the rhythm until a slow count of five or six is possible, with a pause of two or three between in and out breathing.*

6. *Be ready. Self-confidence comes a lot easier to the person who is sure their ready.*

7. *Strengthen your inner self. Write down in your journal what you will be able to do once you* *acquire the confidence you need.*

8. *Begin changing what you do not like about yourself. Confidence comes from within. You need to concentrate on the positive things about yourself.*

9. *Notice the change in our self-esteem and self-confidence. Reward yourself each time you do something that makes you feel proud...go out some place or take it easy for the day.*

10. *Learn how to give and take. Confidence is being able to find a balance between giving help to people and excepting when we need help.*

11. *Have a tremendous amount of pride in yourself. Remember, you are number one!*

EPILEPSY

The Most Important Secrets
You Must Learn
In Order To Live, Learn, and Be Happy With Epilepsy

Chapter 13: Worried, Lost, Confused?

Do you ever feel **unhappy with yourself**? Do you **worry** that you are never going to get where you want to in life? Would you like to **improve your self-self, your self-image and feel like you have some direction in your life?** If you want to look in the mirror and like the person who you see, look at yourself with respect, then this may be the most important article you are ever going to read.

Changing yourself mentally, physically and spiritually does not happen overnight. The process of change takes much time and energy, so do not get discouraged. While focusing on this seven-step plan you will begin to see the changes in yourself as they start to occur. I felt exceptionally proud of myself when I saw myself begin to change for the better. My self-esteem improved and I no longer cared what others thought about me. For the first time I was concerned about what I thought, not what others thought. Once you begin working on this program, you will begin seeing results and realize that this program is worth doing.

Remember, you cannot say you want to change; you have really to want it to do it. The motivation to want to change has to come from within your heart. Saying you want to change is easy, but you have really to want it to do it for it to happen. Otherwise, the change will never occur.

Below, are the **seven steps** to the beginning of the **transformation process**: The transformation process begins when you realize it is time to change and you finally develop the stamina to do what it takes to improve yourself? I created seven steps so that starting the program the right way will be easier for you. Your mind, body and soul all have to be on the same track, functioning as one or else the program will not work for you. The most important part of the program is the beginning. You have to have the correct perception of what you will be doing now and where you will be in the years to follow.

I used these seven steps myself to help me change my outlook on epilepsy. I was able love myself and not be ashamed that I was epileptic. I felt capable to live the life that I wanted to. I felt like a different, better person. I strongly believe that if you follow this program it will help to live your life as an epileptic proudly, creatively, and happily.

Below are the seven steps to lead you to a new beginning:

STEP ONE:

PATIENCE - This is the first step to living a happy and healthy life with epilepsy is developing **patience**. You will need patience to work this program successfully. Changing your outlook on epilepsy is going to take time, devotion, and hard work. Succeeding with this program comes by being patient in wanting to see results.

This exercise will help you even if you are already a patient person because it will relax you and increase your motivational skills simultaneously.

1. *Take a hot bubble bath for fifteen minutes. Also, place an oatmeal bath in the water.*

2. *Lie in the bathtub and close your eyes, take four deep breaths slowly.*

3. *While you are taking these deep breaths clear all thoughts from your mind. Focus on the feeling of the warm water touching your body and the breathing techniques that you are doing at that moment.*

4. *Think about something positive and pleasant. Envision something that makes you happy. Focus on something that makes you feel good about yourself.*

5. *Let go of any negative thoughts that you have stored in your mind. Just to think about one thing that makes you feel good about yourself.*

6. *Take four more deep breaths, relax for a minute, and get out of the bathtub.*

7. *Get dressed, go to a quiet place, and sit on the floor. Close your eyes and slowly bend forward, relaxing any tight muscles that are causing you tension. Bend to the left, stretching your arms as far as they will go, then stretch to the right, repeating the movement.*

8. *Take five more deep breaths and say aloud "I have the patience to change myself and become the person I want to become in life."*

Say, "I have epilepsy and there is nothing wrong with being an epileptic."

9. *Repeat step seven and eight*

10. *Take five more deep breaths and listen to yourself when you are doing this exercise. Concentrate on yourself while doing this exercise. Do not let any distractions impose on your quiet time. Do not think about anything except this exercise and the techniques it involves*

You need to change your outlook on epilepsy by not letting your epilepsy take control of you. As I was growing up, I always made believe that I did not have epilepsy. By doing this, I was only hurting myself. Accepting epilepsy into my life has helped me tremendously. I have released much of the anger that I held inside myself, and have focused on other parts of my life; as a result, I have become a stronger person, extremely proud of the person I have become. You need to do the same. It may take time to get to this point. That is why you need to have patience to come to believe you can do anything you put your mind too. Thinking positive thoughts about yourself will help you get a long **way in life.**

STEP TWO:

Step 2 teaches you how to recognize all the wonderful things about yourself. Judging other people is very easy. Looking at ourselves honestly, however, is difficult. Sometimes we do not focus on ourselves because we have become so preoccupied focusing on everyone else that we forget number one. This step helps recognize all the good things about you. You will begin to have a more positive outlook on life. First, you need to ask, "What do I have to change about myself? What parts of my life need to be readjusted? What are my strengths? **What are my weaknesses?**

Before you answer these questions, get yourself a notebook. To document your answers to these questions and keep track of your progress. The journal helps you see your characteristics and change the ones you dislike. Look at the positive things about yourself and commend yourself for the accomplishments that you achieved and work on changing the negative characteristics that we all carry within us. Begin the journal by listing all the positive things about yourself on the first page. Make a list from one to 10; call this list:

"THE POSITIVE STRENGTHS ABOUT ME"

Your positive points are the strengths that will take you through life. Start your journal with these positive characteristics about yourself. Seeing your strengths itemized in your daily journal will give you encouragement. Each day as you open this journal, you will be looking at all the good things about yourself that will give you motivation to make this program achieve your highest potential. On the next page, create a list and write down your weaknesses. Make a list that looks as the following. Remember, be honest with yourself, and make sure you focus not only on your strengths, but also on your weaknesses. Reviewing our weaknesses can help us see more clearly, what has to be changed in your lives. Make a list from one to 10; call this list:

"THINGS ABOUT MYSELF THAT I NEED TO WORK ON"

On the next page, list ten goals you want to do this week to change your outlook on epilepsy and how you feel about yourself. This will help you gain some insight into what you need to start doing for yourself. Start planning what you want the new you to be like. Each time you accomplish a goal, put a star next to it.

Then list ten long-term goals of what you want to have accomplished and where you want to be in a month's time. Focus on how you are going to accept and love yourself. You need to be proud of the person you have transformed in too. How you are going to change the characteristics about yourself that make you unhappy. Call the list:

"CREATING THE NEW ME"

Goals for the Month

Create a page in the beginning of the book called the **priority calendar**. Ask yourself these questions.

Priority Calendar

1. What do you regret not having made more time for? (List ten things)

2. If you had more time, what would you do with it? (List ten things)

3. What are the top ten priorities in your life right now? (List ten things)

4. What are your family-related goals? (List ten things)

5. What are your business goals? (List ten things)

In the back of the journal, take a quarter of the notebook and title it your **Daily Diary**. Dedicate the diary to someone you care about and feel close too, someone who would be proud to see you accepting that you have epilepsy and living the life, you want to lead. Dedicating the journal to someone, you care about gives you motivation to want to change for the better. Write about the goals you have accomplished and explain how it made you feel to reach them. Write about these achievements are making you into a better person and how their helping you with your epilepsy. Describe what you had to do to achieve the goals.

STEP THREE:

Step 3 is about the importance of how to develop self-determination. Self-determination requires that you make an agreement with yourself and keep it. You must have faith in yourself that you are going to do anything that you put your mind to. Your motivation to accomplish this program will become easier as each day goes by. Saying you are going to improve is easy, but you cannot just say you are going to improve; you have to get out and accomplish the goals that you have set for yourself. You cannot accomplish too many goals in one day. Changing takes time and as step one says, "You need to have patience." Try to accomplish one goal a week at first. Maybe two goals, if you have the time. Then accomplish another goal each week until you get to ten goals. These goals do not have to be difficult. You can set several little goals maybe only one large one. Working on yourself can be tough if you have a busy schedule; nevertheless, do not let that stop you. You have to make time for yourself. Remember; you come first in life. You need to believe that you are the best. You cannot take care of the people who mean the most to you or do the things in life that you want to do, if you are mental, physical and spiritual well-being is not intact and strong.

STEP FOUR:

Reward yourself every time you achieve a goal. Your achievements are important and you should not treat them lightly. For example, take in a movie or reward yourself with some quiet time to relax and focus just on yourself. To me there is nothing better than having sometime alone with you. Do something that makes you happy. Remember, you cannot make others happy until you are happy with yourself.

STEP FIVE:

In your journal, make a list called **Record of Successes** and itemize all the achievements that you have accomplished. Create a list of everything good you believe you have done for yourself. This will make you feel good about yourself.

For example, include the following:

RECORD OF SUCCESSES THROUGHOUT MY LIFE

My Achievements

(List ten things)

RECORD OF SUCCESSES DURING THIS PROGRAM

My Achievements

(List ten things)

STEP SIX:

Develop a special time in the day for quiet time. Studies have shown that individuals who have a daily quiet time are less likely to become ill, and heal faster from illness than those who do not. Take a few minutes during the day to write in your journal. Try to make it the same time each day. Perhaps when you are by yourself, maybe in the morning or just before, you start the day. You could also wait until everyone goes to sleep so that no one will bother you. Give yourself at least fifteen minutes to a half hour. Relax, and while you are writing and relaxing ask yourself the question;

"Where am I headed in life and where do I want to be a year from now?" Then write about it. Make sure you are focusing on the things you want to accomplish in life.

The only way you will succeed in this life is making sure that epilepsy does not control your life. You need to feel proud of the changes you are making with this program. Focus on what you have accomplished. Think about how you feel about having epilepsy while writing in this journal. The goal is to let yourself open up and write intimately and honestly about how you feel. This method helps heal your wounds so you can get on with your life. You have to learn to understand why you have reacted the way you have about having epilepsy. Think of ways to strengthen yourself spiritually and emotionally. Make sure you do not limit yourself because you feel sorry for yourself because you have epilepsy. That is self-pity and it is unhealthy. You will never get anywhere in life if you pity yourself. Free yourself from any walls you have built around yourself. This program will help you do that. Become the person you want to be.

STEP SEVEN:

Now repeat the 7-step process each day. Once you complete the seven steps go back and review the things you have written in your journal. These steps are a new way of life. Keep doing these seven steps each day until you get to where you want to be and you have become completely satisfied with yourself. There is so much knowledge out there for us to learn. It is there for anyone who wants it. I always add more goals to my list. You should always work on bettering yourself. Everyone is special everyone has something unique about them. It is your job to find out what those unique qualities are in you and how to make them work for YOU.

SECTION 3: KEEPING YOURSELF IN GOOD HEALTH CAN HELP YOUR EPILEPSY

Learn how taking care of yourself can help your epilepsy.

EPILEPSY

The Most Important Secrets
You Must Learn
In Order To Live, Learn, and Be Happy With Epilepsy

Chapter 14: How Keeping Yourself in Good Health Can Help Your Epilepsy

To keep your seizures under control, you cannot just pop a pill in your mouth and think that is all there is to it. You need to keep yourself healthy by eating right, exercising, and sleeping properly. These are important factors in helping to control your epilepsy. Certainly, the medication we take to control our seizures plays an important role in our lives, yet if we do not take care of our bodies, we could cause ourselves to have seizures.

We need to try to take the best possible care of ourselves emotionally, spiritually and physically. The way we take care of ourselves affects us in many ways. When I started to take care of myself physically, I began to notice an increase in my energy level. Before then I was lacking energy, feeling fatigue, and feeling sleepy. When I started to focus on my health and began to take better care of myself, I felt more emotionally stable and spiritually in touch with myself. I understood myself even better than when I began the program. I felt so proud of myself. Physically I looked like a new person and spiritually I felt like a new person. My self -esteem rose immensely. By feeling and looking healthy you begin to see yourself as a desirable individual about whom who you can feel proud. Once you start to feel proud of yourself, you will begin to feel like no task is too hard for you to achieve.

I began to realize the importance of keeping myself in good health. In addition, how my health was affecting my epilepsy when I went to see the neurologist one day. I was feeling fatigue and sluggish. My doctor told me that I was overweight for my height and built. I was twenty-two, five foot two, and 146 pounds. The neurologist said, if I lost the weight I would feel better about myself both mentally and physically. By losing the weight, I would help overcome my fatigue and help my epilepsy. My doctor had also mentioned that if I lost weight, most likely the doses of medicine I was taking to control my seizures would decrease. Personally, I was unhappy with my physical appearance. At this point, I realized that I could no longer let myself get any heavier.

The first thing I did to help myself get physically back into shape was change my eating habits. If I were going to lose weight and get back into shape, I needed to change the way I consumed food. I love to eat, just like most people. I had an appetite for fatty foods, sweets, ham and eggs, cream cheese and other good foods that are not so good for the body. Pay attention to what you are eating and eliminate any foods that are unhealthy and do not agree with your body. Everybody's metabolism is different, so you need to eat healthy foods that work best for you. For example, my body reacts well to carbohydrates whereas my husband's body retains much water if he ingests too many carbohydrates into his system.

I began eating mostly proteins and eliminated fatty foods as best I could. You need some fats in your meal plan, but make sure they are the right types of fats. Be careful with the fat-free foods they have on the market. They may have zero fat grams, but the amount of calories could be just as bad as a fatty food with many fat grams in it. To make the fat-free foods taste good they use a lot of sugar, which causes you to gain weight.

If you are at a weight that you are content with, then you

should continue to eat healthy to maintain that weight and to keep in shape. Everything you put in your body affects your epilepsy. You should try to stabilize the amount of calories you consume each day after you decide the amount of calories you want to eat.

I stopped eating any cheeses unless they were fat-free. I stopped using any types of breads or other food products that had a high fat content. The bad thing about bread is that, breads make you feel full. When it goes into your body, it turns into sugar and increase your appetite. Soon you become hungry and want to eat again. Chinese food is high in carbohydrates, and has the same effect. The calories keep adding up and the salt in the food makes you extremely bloated. You should read the food labels when you shop in a grocery store. I ate many fat-free foods that were low in calories. I made sure that the product I was buying had little sugar and sodium in it. This was step number one to getting me back on the right track. I tried to eat as few fat grams as possible in a day.

Drinking water is an important step to eating healthy. I

made sure I drank as much water as my body could consume, which is important. You are trying to get your body back into shape and lose weight Water helps you flush all those unwanted impurities out of your system. The human body contains fifty to 70% water. Because water does not remain stored in the body, we must replace it continually. Water contains no fat grams or calories and is one of the healthiest fluids to drink. Adults must consume two to three liters of some form of liquid each day.

When I became hungry during the day, I made sure I ate healthy snacks such as bananas and yogurt. I cut out all the bad foods, such as the chips, ice cream, cakes, etc. I would have healthy meats such as chicken and turkey. Meats contain many valuable nutrients among them is protein. Nevertheless, be careful also because meat also contains cholesterol.

I cut down on the mayonnaise, ketchup and all the foods that put on weight and hold water. It helped also when I ate slowly. By eating slowly, I would enjoy the meal more and not eat as much. I made sure I also ate breakfast in the morning. I noticed that when I did not eat breakfast, I would eat more during the day or at dinnertime. You want to avoid eating big dinners because the food lies on your stomach later in the evening and you do not burn as many calories. The food just lies there in your stomach. My daily diet consisted of a healthy breakfast, snack, lunch and dinner and a light snack. I felt fulfilled and I lost thirty pounds!

Losing the weight helped me because I was able decrease the medication I was taking. After I began to lose weight, my urge to eat decreased and I felt better about myself emotionally and physically. My body was feeling better and spiritually I felt my inner self began to feel at peace.

The Different Food Groups

- ✓ Fat Oils and Sweets
- ✓ Milk, Yogurt and the Cheese Group
- ✓ Dry Beans, Eggs, and Nut Group

- ✓ Vegetables and Fruit Group Starches, Grains, Pasta, Rice, Bread and Cereal

The bread cereal group includes all breads and cereals that are whole-grain, enriched, or restored. All cereals are very high in starch, and they are good, generally inexpensive sources of energy. The fat content of cereal products generally is very low unless the germ is included. Whole-grain products contribute significant quantities of fiber and such trace vitamins and minerals as antithetic acid, vitamin E, zinc, copper, manganese, and molybdenum.

Most vegetables are important sources of minerals, vitamins, and cellulose. Certain vegetables, such as potatoes, contribute appreciable quantities of starch. Large amounts of the minerals calcium and iron are in vegetables, particularly beans, peas, and broccoli. Vegetables also help meet the body's need for sodium, chloride, cobalt, copper, magnesium, manganese, phosphorus, and potassium. Carotenes (the precursor of vitamin A) and ascorbic acid (vitamin C) are abundant in many vegetables. Vegetables are useful as sources of roughage.

The nutritional value of fruits varies. Some fruits are composed largely of water, but contain valuable vitamins. The citrus fruits are a valuable source of vitamin C, and yellow colored fruits, such as peaches, contain carotene. Dried fruits contain an ample amount of iron, and figs and oranges are an excellent source of calcium. Like vegetables, fruits have high cellulose content.

The milk group includes milk and milk products, cheese, and ice cream. Milk is a complete protein food containing several protein complexes. It also contains important amounts of most nutrients, but it is very low in iron and ascorbic acid and low in niacin. Calcium and phosphorus levels in milk are very high. Vitamin A levels are high in whole milk, but this fat-soluble vitamin is removed in the production of skim milk. Riboflavin is present in significant quantities in milk unless the milk has been exposed to light.

The meat and meat substitutes group includes beef; veal; lamb; pork; organ meats such as liver, heart, and kidney; poultry and eggs; fish and shellfish; and dried peas, beans, and nuts. The meat group contains many valuable nutrients. One of its main nutrients is protein, but meat also contains cholesterol, which is believed to contribute to coronary artery disease. The minerals copper, iron, and phosphorus occur in meats in significant amounts, particularly iron and copper in liver. Different meats vary in their vitamin content. Liver usually contains a useful amount of vitamin A. Thiamine, riboflavin, and niacin, all B vitamins, occur in significant amounts in all meats.

Other Foods such as, butter, margarine, other fats, oils, sugars, or unenriched refined-grain products are included in the diet to round out meals and satisfy the appetite. Fats, oils, and sugars are added to other foods during preparation of the meal or at the table. These foods supply calories and can add to total nutrients in meals.

For many years, the United States Department of Agriculture (USDA) issued dietary guidelines based on four basic food groups--meat and meat substitutes, fruits and vegetables, milk and dairy products, and grains, including bread and cereals and a balanced diet would include at least one food from each group in each meal every day. In 1980 the U.S. Department of Health and Human Services recommended that people eat a variety of foods daily, including fruits; vegetables; whole and enriched grain products; dairy products; meats, poultry, fish, and eggs; and dried peas and beans. While recognizing that certain people (for example, pregnant women, the elderly, and infants) have special nutritional needs, the report stressed that for most people the greater the variety of foods eaten, the less likely is a deficiency or excess of any single nutrient to develop.

The report emphasized that people should increase their consumption of complex carbohydrates fruits, vegetables, and other unrefined foods and naturally occurring sugars. It also recommended reducing the consumption of refined and processed sugars. It encouraged a reduction in fat consumption by decreasing the amount of fatty meats and replacing foods that have saturated fats with those having unsaturated fats. A reduction in the sodium intake by decreasing the amount of salt added to food was also recommended.

Research findings on nutrition, in the USDA and the Department of Health and Human Services changed the daily diet recommendations from the square of the four food groups to a food pyramid, with foods that should be eaten more often at the base, and those used less frequently at the top. The emphasis is on consuming less of the group's meat and meat substitutes, dairy products, and oils and fats, and more of the breads and cereals, and fruits and vegetables. When properly followed the food pyramid teaches the use of a wide variety of food items, moderation in total food intake, and proportionality among the food groups to ensure adequate nutrient intake.

Vitamins are carbon-containing substances that are required for normal metabolism but are not synthesized in the body. They are found from outside sources as food and water or are administered orally or intravenously. Exceptions to this definition include vitamin D, which is synthesized in the body to a limited extent, and vitamins B (12) and K, which are synthesized by bacterial flora in the intestinal tract. Minerals also must be obtained from outside sources. Minerals–such as calcium, iodine, and iron–are an essential part of all cells and body fluids and enter many functions.

Vitamins and minerals function as "cofactors" in the metabolism of products in the body. Most aspects of bodily metabolism proceed with the aid of specific enzymes, but if additional catalysts were not present for example, the cofactor vitamins, and minerals the reactions would proceed so slowly that they would be ineffective.
Vitamin A has many important functions in the body that relate to membrane integrity, especially of epithelial cells and mucous membranes. It is also essential for bone growth, reproduction, and embryonic development.

Vitamin D primarily regulates calcium metabolism by determining the movement of calcium from intestines to blood and from blood to bone. It interacts with parathyroid hormone and calcitonin (see hormone, animal) in controlling calcium levels. Thus, vitamin D is today more legitimately considered a hormone rather than a vitamin.

Vitamin E is considered to have possible value in decreasing the risk of cancer; it has shown little therapeutic value in other diseases. Fortunately, it is relatively nontoxic. Vitamin K is essential for synthesis by the liver of several factors necessary for the clotting of blood. A wide variety of vegetables, egg yolk, liver, and fish oils contain this vitamin.

With the exception of vitamin C (ascorbic acid), water-soluble vitamins belong mainly to what has been termed the B complex of vitamins. The better-known B vitamins are thiamine (B (1)), riboflavin (B (2)), niacin (B (3)), pyridoxine (B (6)), pantothenic acid, lecithin, choline, inositol, and paraaminobenzoic acid (PABA). Two other members are folic acid and cyanocobalamin (B (12)). Yeast and liver are natural sources of most of these vitamins.

Thiamine, the first B vitamin functions as a coenzyme in the form of thiamine pyrophosphate and is important in carbohydrate intermediary metabolism. Riboflavin (B (2) serves as coenzymes for a wide variety of respiratory proteins (see metabolism). Vitamin B (6), functions in human metabolism in the conversion processes of amino acids, including decarboxylation, transamination, and racemization.

In the body folic acid is converted to folinic acid (5-formyl-tetrahydrofolic acid), the coenzyme form, which accepts 1-carbon units important in the metabolism of many body compounds. Nucleic acid synthesis cannot take place without the presence of folic acid. Vitamin B (12), almost all organisms need this vitamin but only in very small amounts.

For vitamin C, a sufficient daily intake of fresh orange juice provides enough of the vitamin for most purposes. The body's requirements for calcium are generally met by eating or drinking dairy products, especially milk. Most calcium (90 percent) is stored in bone, with a constant exchange occurring among blood, tissue, and bone.

Iron is a vital component of hemoglobin and of certain respiratory enzymes. Foods high in iron content include meat (liver and heart), egg yolk, wheat germ, and most green vegetables. The average diet contains 10 to 15 mg a day, adequate for most people.

Magnesium is an essential element in human metabolism and functions in the activities of muscles and nerves, protein synthesis, and many other reactions. Fluorine as fluoride is a requirement to bind calcium in bones. Micro amounts of such elements as boron, chromium, chlorine, copper, manganese, molybdenum, selenium, silicon, sulfur, and vanadium are considered necessary to health.

Normal diets appear to provide adequate amounts of trace minerals, but effects such as the linking of high levels of fructose in the diet with copper deficiency problems are the subject of ongoing research. Vitamins and minerals are an important factor to keeping yourself healthy.

When I started incorporating vitamins into my daily schedule, I began noticing a change in the way I felt physically. When I started eating healthily, I started using a variety of vitamins and herbs that were supposed to be helpful for epilepsy disorders. The vitamins I used were L-Taurine L-tyrosine(amino acids), vitamin B6 and B12, calcium and folic acid. They called an herb that I also had tried a black cohosh. Black cohosh is an eastern North American perennial herb Cimicifuga racemosa. It has a powerful action as a relaxant and a normalizer of the female reproductive system. It may be used in cases of painful or delayed menstruation, ovarian cramps, or cramping pain in the womb. It has a normalizing action on the balance of female sex hormones and may be used safely to regain normal hormonal activity. Used often for the treatment of neurological pain. As a relaxing nervine, it may be used in many situations where such an agent is needed. Medical research has not proven that these vitamins and herbs stop seizures. Nevertheless, they have been used for decades and are said to be beneficial. I still have the same amount of seizures (two to three a month), but I have noticed an increase in my energy. I also had a couple of incidents when I was about to go into an aura. For the first time, I could work with my body and stop myself from having an aura or petite mal seizure by using relaxation exercises. I took deep breaths and slowly released the air through my mouth as I thought positive thoughts.

I have recently incorporated an extra iron vitamin into my diet. Remember, if you decide to use vitamins as a part of your daily diet than it is always safe to first discuss it with your doctor. I believe exercise can help control seizures. Epilepsy does not stop you from being athletic and keeping your body in shape. Some of the greatest athletes had epilepsy. French cyclist Marion Clignet won a silver medal in the 1996 Olympics. Hal Lanier, a former shortstop with the San Francisco Giant; Greg Walker, a former first baseman with the Chicago White Sox, and Buddy Bell, who played seventeen seasons of professional baseball before retiring in 1988, all reportedly had epilepsy as did basketball player Bobby Jones, who played for the Denver Nuggets and Philadelphia '76ers. Exercise helps to build or maintain strength and endurance and to make the body healthier. Exercising is good because it also helps you spiritually. Exercise has both physical and psychological benefits. Regular exercise helps develop muscle tone and strength and control weight. Besides strengthening the muscles, including the heart, regular exercise is believed to make bones stronger by increasing calcium uptake.

Exercise would be extremely beneficial for all individuals, especially for epileptics' who have been using Tegretol for many years. The main problems associated with Tegretol are that it causes your bones to ache. Exercise reduces high blood pressure and cholesterol levels.

Psychologically, regular exercise contributes to a feeling of well-being, and relieves stress. It helps you to feel at peace with yourself. In a study reported in the professional journal Epilepsia, conducted at the department of psychology, University of Alabama at Birmingham exercise proved to have a positive effect on the lives of a hundred and thirty-three people with epilepsy. Fifty-four of them were men and seventy-nine of them were women. Individuals who exercise at least three times a week for a minimum of twenty minutes reported fewer problems with depression and stress.

Another study was conducted on how exercise effects epileptics were done at the University of Sport and Physical Education. In these, study fifteen Norwegian women with drug-resistant epilepsy spent fifteen weeks taking exercise classes twice a week for an hour. They combined aerobic dancing with strength training and stretching. The median number of seizures decreased from 2.9 to 1.7 during the experimental exercise phase. The women also had fewer health complaints, such as muscle pains, sleep problems, and fatigue.

People who exercise are regularly more likely to continue exercise throughout their lives. Once I got into the habit of exercising, it no longer became a chore. Exercising became a physical activity that I enjoyed doing in my spare time. I made sure that I scheduled my days so I would at least have three to four days of exercise. Exercising should take place at least every other day for a period of fifteen to sixty minutes. I enjoy walking, floor aerobics, weights and working on the machines. I would work out a little each day and exercise with the TV or the radio. Whatever motivated me then, but remember before you begin any exercise program you should check with your doctor first.

When I exercised, it made me feel good inside. Exercise should be some activity that you can do that would not strain your body. It can be walking, jogging, running, aerobics, bench stepping, hiking, jazzercise, bodybuilding, swimming, dance, or anything that you enjoy. You should do some kind of exercise to keep your body in shape. The older you become the more important is to exercise. Exercise affects the aging body, helping to maintain fitness and slow down the physical effects of aging. If not properly exercised, the aging body can develop problems in the muscles, bones, and cardiovascular system. As you get older, your muscles begin to wither away and lose their tone, leading to more frequent tearing of the tendons in the muscles. Your bones become weak and brittle, fracturing easily and more often. Therefore, it is important that you take responsibility and keep yourself in good shape. Keeping yourself healthy will help you and your epilepsy.

As I was working to accept my epilepsy, I noticed myself changing emotionally. I felt better about myself. I would look in the mirror and be proud of the person I was seeing.

In college, my marketing professor assigned us a book to read. It was the most boring book I ever read in my life. I could feel the stress increase as I kept reading the book. The book had no purpose and I could not understand why he would assign us to read this book. Suddenly I felt an aura start to come on and then the rest of the seizure. My point is that I could have avoided the stress by putting the book the down and reading it at different times. I had created the stress myself. We have the power to control stress and thus help our body avoid seizures. We need to understand how our bodies work and listen to its messages. Because we have epilepsy, we need to be a little more careful than the average individual is.

I never thought when I was

growing up as a child that I would have to be limited in enjoying certain activities I wanted to explore. I thought I could do everything. I realize now, as a young woman, there is no reason for me to lower my expectations, but there is also no reason to push myself over the limit. No one on this earth is a 100% perfect. We all have our faults. I work all the time by trying to make myself into a stronger human being emotionally, physically and spiritually. Working on myself makes me feel like I can fight the battle of epilepsy and so can you!

Everybody has different characteristics that make up their personalities. Epilepsy is just one part of me. I cannot change the fact that I have epilepsy. I have to accept it the fact that I have epilepsy and learn to live with it productively. You will succeed and become a better person, if you think positively and productively. You will feel the strength in yourself to accept your epilepsy. Think about who you and where you are headed in life. It is up to you to make something of yourself.

EPILEPSY

The Most Important Secrets
You Must Learn
In Order To Live, Learn, and Be Happy With Epilepsy

Glossary

EPILEPSY GLOSSARY

PROVIDING KNOWLEDGE FOR A HEALTHIER YOU

Getting familiar with these terms will help discussions with your doctors

Absence seizure: (also know as "dialeptic seizure" or "petit mal seizure") a seizure that causes a brief loss of awareness. During an absence seizure, the patient stops any activity and stares blankly. Rarely, there may be some blinking.

Accommodation (or Reasonable accommodation): Any change in the work environment or in the way things are customarily done that enables an individual with a disability to have equal employment opportunities.

ADHD: Attention deficit hyperactivity disorder.

Adjunct: Something added to another thing in a subordinate position or use; for example, an adjunct drug is one used in addition to another drug, not alone *(add-on therapy)*.

Adverse effects: side effect; negative effect from medication or therapy

Ambulatory EEG monitoring: A system for recording the electroencephalogram for a prolonged period (typically 18 to 24 hours) in an outpatient; the electrodes are connected to a small cassette tape recorder.

Americans with Disabilities Act: A law that makes discrimination against people with disabilities illegal; the act applies to employment, access to public places, and places of accommodation.

Anhedonia: Failure to enjoy positive emotional experiences

Anterograde memory: The ability to form new memories; memories for events occurring after a problem such as a head trauma or seizure.

Anticholinergic: Drugs that block the cholinergic neurotransmitter system

Antiepileptic drug: A medication used to control both convulsive and nonconvulsive seizures; sometimes called an *anticonvulsant*.

Apathy: Lack of interest and motivation

Anticonvulsant: an antiepileptic drug used to control both convulsive and nonconvulsive seizures.
Atonic seizure: a seizure that causes a sudden loss of muscle tone, particularly in the arms and legs, and often causes the patient to fall.
Attention deficit disorder (ADD): An impairment in the ability to focus or maintain attention.

Atypical absence seizure: A staring spell characterized by partial impairment of consciousness; often occurs in children with the Lennox-Gastaut syndrome; the EEG shows slow (less than 3 per second) spike-and-wave discharges.

Aura: a warning or initial symptom at the beginning of a seizure, experienced by the patient, but not visible to observers. Auras may progress to become focal or even generalized seizures, or they may exist alone.

Autoinduction (of metabolism): A process in which continued administration of a drug leads to an increase in the rate at which the drug is metabolized.

Automatism: Automatic, involuntary movement during a seizure; may involve mouth, hand, leg, or body movements; consciousness is usually impaired; occurs during complex partial and absence seizures and after tonic-clonic seizures.

Autonomic: Pertaining to the autonomic nervous system, which controls bodily functions that are not under conscious control (e.g., heartbeat, breathing, sweating); some partial seizures may cause only autonomic symptoms; changes in autonomic functions are common during many seizures.

Autosomal dominant: A mode of inheritance in which a gene is passed on by either parent; in most cases, the child has a 50% chance of inheriting the gene; the *expression* of the gene (that is, the development of the physical trait or the disorder) can vary considerably among different individuals with the same gene.

Autosomal recessive: A mode of inheritance in which an individual has two copies of a gene that requires both copies for *expression*, or development, of the trait. Both parents must be *carriers* (that is, they have only one copy of the gene and, therefore, do not have the physical trait that the gene confers) or have the trait (that is, have two copies of the same gene).

Axon: The part of the nerve cell (neuron) that communicates with other cells, similar to a telephone wire; the axon is often covered with myelin, an insulating fatty layer, which functions similarly to plastic around a copper wire.

Benign: Favorable for recovery.

Benign rolandic epilepsy: An epilepsy syndrome of childhood characterized by partial seizures occuring at night and often involving the face and tongue; the seizures may progress to *tonic-clonic* seizures, have a characteristic EEG pattern, are easily controlled with medications but may not require treatment, and are outgrown by age 16 years.

Blood drug level: The concentration, or amount, of circulating drug in the bloodstream, measured in micrograms (µg) or nanograms (ng) per milliliter (mL). The concentration may be measured as the free or total level because some of the drug is bound to the protein in the blood and some is not; the free level is the amount of drug that is "free" (unbound); the *total level* is the amount of drug that is both bound and unbound to the blood protein; the drug that is free (unbound) is the portion that reaches the brain and exerts an effect on the disorder.

Brand-name drug: Medication manufactured by a major pharmaceutical company; the drugs are often expensive, but tend to be uniform in the amount of drug and the method of preparation.

Breath-holding spells: Episodes in children in which intense crying or an emotional upset is followed by interruption of breathing and sometimes loss of consciousness; the episodes are not harmful, but when prolonged, slight jerking movements may occur.

Catamenial: Referring to the menses or to menstruation; with regard to women with epilepsy, a tendency for seizures to occur around the time of the menses.

Cerebral hemisphere: One side of the cerebrum (upper brain); each hemisphere contains four lobes (frontal, parietal, occipital, and temporal).

Cognitive: Pertaining to the mental processes of perceiving, thinking, and remembering; used loosely to refer to intellectual functions as opposed to physical functions.

Comorbid: A disorder that is present in association with another

Clonic seizure: Repetitive, rhythmic jerks that involve all or part of the body.

Comprehensive epilepsy centers : Comprehensive epilepsy centers are clinics staffed by epileptologists and other experts in epilepsy treatment. They are valuable resources for anyone who has unresolved problems related to definite or suspected epilepsy. Patients may be referred to a comprehensive epilepsy center for a single outpatient visit to assess their diagnosis and therapy, or they may receive long-term follow-up and treatment, including epilepsy surgery or the use of new medications that are still being investigated.

Computed tomography (CT): A scanning technique that uses x-rays and computers to create pictures of the inside of the body; shows the structure of the brain; not as sensitive as MRI.

Consciousness: State of awareness; if consciousness is preserved during a seizure, the person can respond (either in words or actions, such as raising a hand on command) and recall what occured during the spell.

Controlled study: An experiment in which two groups are the same except that only one receives the drug, treatment, etc. being tested.

Complex partial seizure: a seizure that includes impairment of awareness, for example, patients seem to be "out of it" or "staring into space." Unintentional movements or other movements are frequently a part of the seizure.

Convulsion: An older term for a tonic-clonic seizure.

Convulsive syncope: A fainting episode in which the brain does not receive enough blood, causing a seizure; the episode is not an epileptic seizure, but a result of the faint.

Corpus callosotomy: A surgical technique that disconnects the cerebral hemispheres and is most effective in reducing atonic and tonic-clonic seizures.

Cortical dysplasia: An abnormality in the development and organization of the cerebral cortex that can cause seizures and other neurologic disorders. These disorders can result from abnormal migration of nerve cells during development or can occur with disorders such as tuberous sclerosis or Sturge-Weber syndrome.

Corpus callosum: a band of nerve fibers located deep in the brain that connects the two halves (hemispheres) of the brain. The corpus callosum helps the hemispheres share information.

Corpus callosotomy: an operation that cuts the corpus callosum and interrupts the spread of seizures from one hemisphere of the brain to the other. Callosotomies may be complete, or may involve only a portion of the corpus callosum. Although seizures generally do not completely stop after this procedure, they usually become less severe.

Daily dose: The average amount of medication taken over the course of the day to achieve a therapeutic blood level of the drug, usually measured in milligrams (mg) per kilogram (kg) of the patient's body weight (1 kg = 2.2 pounds).

Deficit: A lack or deficiency of an essential quality or element; for example, a neurologic deficit is a defect in the structure or function of the brain.

Deja vu: Feeling as if one has lived through or experienced this moment before; may occur in people without any medical problems or immediately before a seizure (i.e., as a simple partial seizure).

Development: The process of physical growth and the attainment of intelligence and problem-solving ability that begins in infancy; any interruption of this process by a disease or disorder is called *developmental delay*.

Dose-related effects: Adverse effects that are more likely to occur at times of peak blood levels of a drug.

EEOC: Equal Employment Opportunity Commission.

EF: Epilepsy Foundation.

Efficacy: Effectiveness

Elimination: The removal of waste products from the body.

Encephalitis: An inflammation of the brain, usually caused by a virus.

Epilepsia partialis continua: A continuous or prolonged partial seizure that causes contraction of the muscles; usually restricted to the muscles of the face, arm, or leg; usually not associated with impairment of consciousness.

Epilepsy syndrome: A disorder defined by seizure type, age of onset, clinical and EEG findings, family history, response to therapy, and prognosis.

Epileptiform: Resembling epilepsy or its manifestations; may refer to a pattern on the EEG associated with an increased risk of seizures.

Epileptogenesis: The process(es) that lead to the development of epilepsy

Epileptogenic: Causing epilepsy.

Epileptologist: A neurologist with specialty training in epilepsy.
Excitatory: Stimulating or increasing brain electrical activity; causing nerve cells to fire.

EEG-video monitoring: Continuous simultaneous recording of brainwaves and video observation of the behavior accompanying the EEG. This technique, carried out at comprehensive epilepsy centers, is employed to diagnose epilepsy and localize the seizure focus. The results are useful to determine therapy -- medical or surgical.

Epilepsy: a chronic medical condition marked by recurrent epileptic seizures. Patients may have single seizures as a result of fever, medication withdrawal, etc., but are not labeled as having epilepsy if seizures do not recur.

Epileptogenic zone: the region of the brain responsible for the abnormal electrical signals that cause seizures.

Electrode: a conductive disk (usually metal) attached to the scalp which conveys the electrical activity of the brain through a wire to an EEG machine. During an electroencephalogram, typically around 20 electrodes are temporarily pasted to the scalp.

Electroencephalogram (EEG): a diagnostic test that measures brainwaves, the electrical impulses in the cerebral cortex. This test helps a doctor to diagnose epilepsy.

Epilepsy surgery: a neuro-surgical procedure to prevent further seizures, usually accomplished by resecting the epileptogenic zone. Successful in eliminating seizures in a large majority of patients, depending on the type of epilepsy identified during EEG-video monitoring.

Extratemporal cortical resection: an operation to cut out (resect) brain tissue that contains a seizure focus. "Extratemporal" means the tissue is located in an area of the brain other than the temporal lobe, most often the frontal lobe.

Functional hemispherectomy: a procedure in which portions of one hemisphere of the brain which is not functioning normally are removed, and the corpus callosum is split. This interrupts the communications among the various lobes and between the two hemispheres and prevents the spread of seizures.

Febrile seizure: A seizure associated with high fever in children aged 3 months to 5 years, usually a tonic-clonic seizure; benign in most cases.

Fit: An older term for a seizure, usually a tonic-clonic seizure; still used in some places.

Flattened affect: Decreased emotional range.

Focal seizure: An older term for a partial seizure.

Focus: The center or region of the brain from which seizures begin; used in reference to partial seizures.

Frontal lobe seizure foci: A partial seizure arising in the frontal lobe area of the brain.

Hemisphere: one half of the cerebrum, the largest part of the brain.

Generalized seizure: a seizure that occurs all through the brain.

Generic drug: A drug that is not sold under a brand name; for example, carbamazepine can be obtained as a generic drug or as Tegretol or Carbatrol, its brand names.

Grand-mal seizure: an older term for a seizure in which the patient loses consciousness and collapses. The patient also has body stiffening and violent jerking, and then often goes into a deep sleep. Also known as a generalized convulsion.

Half-life: The time required for the amount of a drug in the blood to decline to half its original value, measured in hours; a drug with a longer half-life lasts longer in the body and, therefore, generally needs to be taken less often than a drug with a shorter half-life.

Hemispherectomy: A surgical procedure to remove a cerebral hemisphere (one side of the brain); the operation is now often modified to remove a portion of the hemisphere and to disconnect the remaining portions.

Hereditary: Passed from one generation to the next through the genes.

Hydrocephalus: A condition associated with obstruction of the cerebrospinal fluid pathways in the brain and accumulation of excess cerebrospinal fluid within the skull.

Hyperventilation: Increased rate and depth of breathing; may be done during the EEG to increase the chances of finding epileptiform or other abnormal activity.

Hypofunction: Diminished functional activity

Hypsarrhythmia: An abnormal EEG pattern of excessive slow activity and multiple areas of epileptiform activity; associated with infantile spasms.

Ketogenic diet: a treatment for epilepsy intended to maintain the starvation or fasting metabolism for a long period in order to create ketones, byproducts of fat-burning metabolism. Seizures often lessen or disappear during periods of fasting. The diet is very high in fat and low in carbohydrates and is most often recommended for children ages 2 through 12 who have been diagnosed with a generalized type of epilepsy, and who have failed to respond to a variety of medications.

Lesionectomy: surgery to remove isolated brain lesions that are responsible for seizure activity.

Lobe: one of the sections of the cerebrum, the largest part of the brain. The lobes are divided into four paired sections (frontal, parietal, occipital, and temporal). The seizure focus is usually located in one of the lobes.

Lumbar puncture: a diagnostic procedure in which the fluid surrounding the spinal cord (cerebrospinal fluid) is withdrawn through a needle and examined in a laboratory. Also known as a spinal tap.

Multiple subpial transection: a surgical procedure to help control seizures that begin in areas of the brain that cannot be safely removed (areas that control movements or speech). The surgeon makes a series of shallow cuts (transections) in the brain tissue to interrupt the movement of seizure impulses.

Myoclonic seizure: a seizure that consists of sporadic jerks, usually on both sides of the body. Patients with these seizures may drop or involuntarily throw objects.

Neurologist: a doctor who specializes in the treatment of epilepsy and other disorders of the brain and nervous system.

Neruron: a single nerve cell. The brain is made up of billions of neurons. Many neurons malfunctioning together are necessary to produce a seizure.

Nonepileptic event: an event that resembles a seizure but is actually produced by another condition, such as Tourette syndrome or heart rhythm disturbances (arrhythmias). Certain psychological conditions can also bring on a nonepileptic event.

Partial seizure: (also know as a "focal seizure") a seizure that occurs in a limited area in only one hemisphere of the brain. This type of seizure is more amenable to treatment with surgery than are generalized seizures.

Ring Chromosome 20 Syndrome: Ring chromosome 20 syndrome (RC20) is one of a number of chromosomal disorders associated with refractory epilepsy. A ring chromosome is formed by the fusion of two arms of a chromosome during pre-natal development. Epilepsy appears to be the first and major clinical symptom of this syndrome, is a constant feature, and is often drug resistant.

Seizure: an event of altered brain function caused by abnormal or excessive electrical discharges in the brain. Most seizures cause sudden changes in behavior or motor function.

Seizure focus: the area of the brain in which a seizure starts.

Status epilepticus: a prolonged seizure (usually defined as lasting longer than 30 minutes) or a series of repeated seizures without regaining consciousness. Status epilepticus is a medical emergency, and medical help should be obtained immediately.

Seizure threshold: Minimal conditions necessary to produce a seizure.

Seizure: A sudden, excessive discharge of nervous-system electrical activity that usually causes a change in behavior.

Selective Serotonin Reuptake Inhibitor (SSRI): A drug that blocks the removal of serotonin from the synapse;

thereby prolonging and increasing the effects of serotonin.

Sensory: Pertaining to the senses (touch, vision, hearing, taste, smell).

Sharp wave: An EEG pattern indicating the potential for epilepsy; "benign" sharp waves are not associated with seizures.

Simple partial seizure: An epileptic seizure that involves only part of the brain and does not impair consciousness.

Single-Photon Emission Computed Tomography (SPECT): A diagnostic test that uses a very low and safe dose of a radioactive compound to measure blood flow in the brain; not as sensitive as PET for baseline (interictal) studies but can more readily be obtained during a seizure.

Slowing: A term used to describe a group of brain waves on the EEG that have a lower frequency than expected for the subject's age and level of alertness and the area of the brain recorded. Slow waves can result from drowsiness or sleep, drugs, or brain injuries and occur during or after seizures.

Social Security Disability Income (SSDI): A federal assistance program for disabled people who have paid

Social Security taxes or are dependents of people who have paid.

Spell: A period, bout, or episode of illness or indisposition; refers to seizures or other disorders that produce brief episodes of behavioral change.

Spike: An EEG pattern strongly correlated with seizures; "benign" spikes are not associated with seizures.

SSA: Social Security Administration

Status epilepticus: A prolonged seizure (usually defined as lasting longer than 30 minutes) or a series of repeated seizures; a continuous state of seizure activity; may occur in almost any seizure type. Status epilepticus is a medical emergency, and medical help should be obtained immediately.

Steady state: A state in which equilibrium has been achieved. In reference to antiepileptic drugs, steady state is achieved when a constant daily dose of a drug produces consistent blood levels of the drug (takes at least five times the half-life of the drug in question).

Structural lesion: Physical abnormality in the brain.

Sturge-Weber syndrome: A disorder of blood vessels affecting the skin of the face, eyes, and brain; brain involvement is associated with seizures.

Supplemental Security Income (SSI): A federal assistance program.

Symptomatic: Referring to a disorder with an identifiable cause; for example, severe head trauma can cause symptomatic epilepsy.

Synapse: The junction between one nerve cell and another nerve cell; the axon of one nerve cell releases a neurotransmitter, which diffuses across the synapse and causes changes in the membrane of the adjacent cell.

Syncope: (pronounced SIN-ko-pee) Fainting.

Syndrome: A group of signs and symptoms that collectively define or characterize a disease or disorder; signs are objective findings such as weakness, and symptoms are subjective findings such as a feeling of fear or tingling in a finger.

Temporal lobe epilepsy: An older term for partial epilepsy arising from the temporal lobe of the brain.

Temporal lobe seizure: A simple or complex partial seizure arising from the temporal lobe of the brain.

Therapeutic blood level: The amount of drug circulating in the bloodstream that brings about seizure control without troublesome adverse effects in most patients. "Subtherapeutic" (lower) levels are effective in some patients, and "supratherapeutic" or "toxic" (higher) levels are tolerated by others.

Threshold: The level at which an event or change occurs

Tic: Repeated involuntary contractions of muscles, such as rapid head jerks or eye blinks, as in Tourette's syndrome; may be under partial voluntary control (for example, can be temporarily suppressed); nonepileptic.

Time to Peak Blood Level: The interval between the time a drug is taken and the time it reaches the highest concentration in the blood.

Todd's paralysis: Weakness after a seizure; originally used to describe muscle weakness on the side of the body opposite the side in which the seizure

Tolerance: Decreased sensitivity to the effects of a substance such as a medication.

Temporal lobe resection: a surgical procedure in which brain tissue in the temporal lobe is cut away (resected) to remove the seizure focus.

Tonic seizure: a seizure that is characterized by stiffening of the muscles, sustained for more than a few seconds.

Tonic-clonic seizure: a seizure marked by loss of consciousness, falling, stiffening, and jerking. This is the hallmark of a generalized motor seizure, which used to be called a "grand mal seizure."

Trauma: An injury or wound caused by external force or violence.

Tuberous sclerosis: A disease in which benign tumors affect the brain, eyes, skin, and internal organs; associated with mental retardation and seizures; inherited as an autosomal dominant trait.

Vagus nerve stimulation: a surgical treatment for epilepsy involving implantation in the neck of an electrode on the vagus nerve. The electrode is connected to a pacemaker that is placed under the skin in the chest. While the VNS is usually programmed to cycle continuously, the patient can turn the stimulator on, using a small magnet placed over the pacemaker, if he or she feels a seizure coming on.

Vagus nerve: a small cranial nerve that passes through the neck and is connected to various areas of the brain and other organs in the body, including the stomach, heart, and lungs.

Video-EEG monitoring: A technique for recording the behavior and the EEG of a patient simultaneously; changes in behavior can be correlated with changes in the EEG; useful for making the diagnosis of epilepsy and localizing the seizure focus.

West's syndrome: An epileptic syndrome characterized by infantile spasms, mental retardation, and an abnormal EEG pattern (hypsarrhythmia); begins before 1 year of age.

Resources

There are plenty of resources available to learn more about epilepsy, both online and in print.

If you or someone you know has been diagnosed with epilepsy, you'll no doubt want more details about this disorder. There are plenty of places to turn for information on treatments, controlling seizures, finding specialists and support groups, or new research, both online and off.

Finding an Epilepsy Organization or Foundation

An epilepsy foundation is a nonprofit organization dedicated to raising awareness about epilepsy. There are many to choose from, and they may each offer a different, specific goal or focus. Here's some information to help you get started.

Epilepsy Foundation
8301 Professional Place
Landover, MD 20785-7223

postmaster@efa.org
http://www.epilepsyfoundation.org
Tel: 301-459-3700 800-EFA-1000 (332-1000)
Fax: 301-577-2684

This group, which promotes awareness and education about epilepsy, offers a Web site with helpful epilepsy information, including:

- ✓ What the disorder is like
- ✓ What seizure symptoms are
- ✓ What it's like to live with epilepsy

The Epilepsy Foundation can help you find your local branch so that you can become involved within your own community. There is also a blog where you can read about others' experiences with epilepsy and write about your own. In addition, the Web site offers updates on current research, and information on what the Epilepsy Foundation is doing to advocate for people with epilepsy.

The American Epilepsy Society

342 North Main Street
West Hartford, CT 06117-2507
(860) 586-7505
This organization was founded to:

- ✓ Encourage epilepsy education and awareness
- ✓ Promote epilepsy research
- ✓ Help cure, treat, and prevent epilepsy

The society raises money for research and offers

continuing education classes for doctors to learn about epilepsy advancements and developments. The group's Web site offers a facts and statistics section, current news about epilepsy, and helpful links to find an epilepsy specialist in your area.

Citizens United for Research in Epilepsy (CURE)

223 W. Erie
Suite 28W
Chicago, IL 60654
info@CUREepilepsy.org
http://www.CUREepilepsy.org
Tel: 312-255-1801 800-765-7118
Fax: 312-255-1809

CURE is an epilepsy group that focuses on raising money to fund research. CURE also works to increase awareness of epilepsy. Its Web site offers educational epilepsy information, as well as a list of fund-raising events that it sponsors.

Epilepsy in Print

Whether you want a reference book to flip through, a guide to strategies for coping with epilepsy, or others' personal accounts of the conditions, there is a book available for you. The Epilepsy Foundation recommends the following:

- ✓ **Epilepsy,** by Orrin Devinsky, MD. A great how-to reference book for individuals with epilepsy and their families. Find information about seizures, medication, and self-care, as well as information on legal and social concerns related to epilepsy.

- **Living Well With Epilepsy,** by Robert J. Gumnit, MD. This book offers information from many epilepsy experts on living with the brain condition and provides perspective on the adjustments that may need to be made to better manage epilepsy.

- **The Ketogenic Diet,** by John M. Freeman, MD, et al. A resource that teaches the details of managing epilepsy and preventing seizures with diet. The ketogenic diet, which focuses on lower carbohydrate and higher fat intake, has been very successful in reducing seizures in children with epilepsy. This book should be used only after consulting with the doctor who is helping to manage your or your child's epilepsy.

- Epilepsy and Pregnancy by Stacey Chillemi and Blanca Vazquez, MD

All of these books are available through the Epilepsy Foundation's Web site and may also at online bookseller sites or in your local bookstore or library.

Child Neurology Foundation - This group is dedicated to education and support for children and families dealing with epilepsy. This link will take you to their web site.

Epilepsy.com - This site features information about epilepsy and offers resources for everyone involved in caring for, or treating, those with epilepsy. This link will take you to their web site.

National Institute of Neurological Disorders and Stroke - This site has information on disorders of the brain and nervous system.

Epilepsy Action
Epilepsy Action aims to improve the quality of life and promote the interests of people living with epilepsy.
http://www.epilepsy.org.uk

Epilepsy Advocate
A detailed epilepsy site for patients and caregivers.
http://www.epilepsyadvocate.com

Epilepsy Bereaved
The leading voluntary organization on Sudden Unexpected Death in Epilepsy (SUDEP) and other epilepsy related deaths.
http://www.sudep.org

Epilepsy.com Information on Epilepsy - including clinical

trial information, treatment and medication details.
http://www.epilepsy.com/

ISAS (Ictal-Interictal SPECT Analysis by SPM)
Website detailing a method of localizing seizure activity using SPECT imaging (ISAS). The analysis is done via a freely available software package (SPM).
http://spect.yale.edu/

STARS
Syncope Trust and Reflex Anoxic Seizures - Complex Syncope often appear just like a 'seizure' or 'fit', this sometimes leaves patients with an incorrect diagnosis of epilepsy.
http://www.stars-us.org

Workshop on Antiepileptic Drug (AED) Monotherapy Indication
A workshop attempt to reach consensus on the best method to obtain FDA approval for monotherapy labeling for antiepileptic drugs (AEDs)
Publicaciones en Español
Crisis Epilépticas: Esperanza en la Investigación

Hope for Hypothalamic Hamartomas (Hope for HH)
P. O. Box 721
Waddell, AZ 85355
admin@hopeforhh.org
http://hopeforhh.org/

Epilepsy Institute
257 Park Avenue South
New York, NY 10010
website@epilepsyinstitute.org
http://www.epilepsyinstitute.org
Tel: 212-677-8550
Fax: 212-677-5825

People Against Childhood Epilepsy (PACE)
7 East 85th Street
Suite A3
New York, NY 10028
pacenyemail@aol.com
http://www.paceusa.org
Tel: 212-665-PACE (7223)
Fax: 212-327-3075

Family Caregiver Alliance/ National Center on Caregiving
180 Montgomery Street
Suite 900
San Francisco, CA 94104
info@caregiver.org
http://www.caregiver.org
Tel: 415-434-3388 800-445-8106
Fax: 415-434-3508

National Council on Patient Information and Education
200-A Monroe Street
Suite 201
Rockville, MD 20850
ncpie@ncpie.info

http://www.talkaboutrx.org
Tel: 301-340-3940
Fax: 301-340-3944

National Family Caregivers Association
10400 Connecticut Avenue
Suite 500
Kensington, MD 20895-3944
info@thefamilycaregiver.org
http://www.thefamilycaregiver.org
Tel: 800-896-3650
Fax: 301-942-2302

National Organization for Rare Disorders (NORD)
P.O. Box 1968
(55 Kenosia Avenue)
Danbury, CT 06813-1968
orphan@rarediseases.org
http://www.rarediseases.org
Tel: 203-744-0100 Voice Mail 800-999-NORD (6673)
Fax: 203-798-2291

International Radio Surgery Association
3002 N. Second Street
Harrisburg, PA 17110
irsa@irsa.org
http://www.irsa.org
Tel: 717-260-9808
Fax: 717-260-9809

Charlie Foundation to Help Cure Pediatric Epilepsy
1223 Wilshire Blvd.
Suite #815

Santa Monica, CA 90403
ketoman@aol.com
http://www.charliefoundation.org
Tel: 310-393-2347
Fax: 310-453-4585

Epilepsy Therapy Project
P.O. Box 742
10. N. Pendleton Street
Middleburg, VA 20118
info@epilepsytherapyproject.org
http://www.epilepsy.com
Tel: 540-687-8077
Fax: 540-687-8066

Antiepileptic Drug Pregnancy Registry
Massachusetts General Hospital
121 Innerbelt Road Room 220
Somerville, MA 02143
info@aedpregnancyregistry.org
http://www2.massgeneral.org/aed/
Tel: 888-AED-AED4 (233-2334)
Fax: 617-724-8307

International Dravet Epilepsy Action League
P.O. Box 797
Deale, MD 20751
info@IDEA-League.org
http://www.IDEA-League.org
Tel: 443-607-8267

Intractable Childhood Epilepsy Alliance
PO Box 365

6360 Shallowford Road
Lewisville, NC 27023
info@ice-epilepsy.org
http://www.ice-epilepsy.org
Tel: 336-946-1570
Fax: 336-946-1571

Epilepsy Websites:

1. **Epilepsy.com** - Epilepsy and seizure diagnosis, treatment, and information for people with epilepsy. **www.epilepsy.com**

2. **www.braintalk.org/**
BrainTalk Communities: Online patient support groups for neurology (from Massachusetts General Hospital) -- a large and active site with separate forums for epilepsy and a very wide range of other disorders.

3. **www.cdc.gov/nccdphp/epilepsy/index.htm**
National Center for Chronic Disease Prevention and Health Promotion: A page dedicated to some general information on epilepsy. The National Center for Chronic Disease Prevention also has a toolkit for

parents of teens with epilepsy:
www.cdc.gov/epilepsy/toolkit/index.htm

4. **www.columbia.edu/**
   a website for The IRELAND Project, based at Columbia University in New York. The IRELAND Project is a multicenter genetic linkage analysis study designed to look for the genes that cause seizures in Rolandic Epilepsy and associated difficulties in reading and speech.

5. **www.emedicinehealth.com/**
   a website specifically designed to address health issues such as epilepsy. The site contains articles written by physicians for patients and consumers.

6. **www.epilepsycenter.com/**
   The Epilepsy Center is an organization that provides

services in northwestern Ohio. Features safety tips, posters for sale, and extensive links.

7. **www.epilepsyfoundation.org/**
   Website of The Epilepsy Foundation, the U.S. national organization that works for people affected by seizures. A good place to get reliable information, especially on legal affairs and community services.

8. **www.epilepsyinstitute.org/**
   the website of The Epilepsy Institute, a social-service organization for people with epilepsy in New York City and Westchester County, New York.

9. **www.epilepsytoronto.org/**
   Website of Epilepsy Toronto, a non-profit organization dedicated to the promotion of independence and optimal quality of life for all people with epilepsy and their families. They

offer a range of epilepsy support services, information programs and education to the public.

10. www.epipro.com

Epilepsy & Brain Mapping Program: A comprehensive healthcare center in Pasadena, California, for treating adult and pediatric epilepsy and other neurological disorders with seizures.

11. http://familydoctor.org/214.xml

Familydoctor.org, a website created by the American Academy of Family Physicians: General information on epilepsy.

12. www.hsc.stonybrook.edu

the website of the Long Island Comprehensive Epilepsy Center.

13. **www.naec-epilepsy.org**

the website of the National Association of Epilepsy Centers. Find a comprehensive epilepsy center near you!

14. **www.neuro.wustl.edu/epilepsy/**

Epilepsy information and local resource centers in St. Loius, MO.

15. **www.nichcy.org/pubs/factshe/fs6txt.htm**

National Dissemination Center for Children with Disabilities: Epilepsy information and resources.

16. **www.nyuepilepsy.org**

Website of the NYU Comprehensive Epilepsy Center, the largest epilepsy center in the United States. The Center offers testing, evaluation, screening, treatment, drug trials, alternative therapies, and surgical intervention for children, adolescents, and adults with all forms of epilepsy.

17. **www.nyuepilepsy.org/cec/**

New York University-Mount Sinai Comprehensive Epilepsy Center website. Information and useful links. The center includes a special program for people with tuberous sclerosis.

18. **www.nyufaces.org**

the website of Finding A Cure for Epilepsy and Seizures (f.a.c.e.s.). f.a.c.e.s. is an epilepsy organization that funds epilepsy research, education and awareness, and community building events for people with epilepsy.

19. **www.paceusa.org/**

Parents Against Childhood Epilepsy is an organization

that provides support for parents of children with epilepsy. Features information regarding fundraising efforts in support of advances in medical research of epilepsy.

20. **http://specialchildren.about.com/cs/epilepsy/**
    A list of epilepsy websites primarily for parents of children with epilepsy.

21. **http://pppl.tblc.org/sesweb/**
    Suncoast Epilepsy Association: A Florida, non-profit organization that provides services to individuals with epilepsy in Pinellas and Pasco Counties.

22. **www.4pep.org**
    The Pediatric Epilepsy Project at UCLA (PEP) was formed with a single, all-important mission: to raise funds, provide financial support and increase community awareness to help sustain the Division of Pediatric Neurology at UCLA.

23. **http://library.thinkquest.org/J001619/**

"Growing Up with Epilepsy" is a ThinkQuest website created by two young, gifted students with their teacher as their coach.

24. **www.bgsm.edu/bgsm/surg-sci/ns/epilepsy.html**

Wake Forest University School of Medicine, Winston-Salem, North Carolina, and Department of Neurosurgery: Contains general facts on epilepsy along with links to numerous websites for the epilepsy community of Wake Forest, as well as, various other sites containing helpful information on epilepsy.

25. **http://faculty.washington.edu/chudler/epi.html**

University of Washington: Professor's website containing epilepsy information and resources.

26. www.bbc.co.uk/health/epilepsy/

A UK website with information on epilepsy and local resources for UK residents living with epilepsy.

27. www.bcepilepsy.com/

The BC Epilepsy Society is a non-profit, charitable organization dedicated to serving the British Columbians living with epilepsy, and their families.

28. www.epilepsiemuseum.de/

A German online museum that features artwork, history, and virtual tours.

29. www.epilepsy.ca/

the website for Epilepsy Canada, featuring information on Canadian resources as well as general facts. Also includes "Kidz Korner" with reassuring explanations for children.

30. **www.epilepsyontario.org/**

Epilepsy Ontario: A website with information on epilepsy and support and volunteer opportunities for the local Ontario epilepsy community.

31. **www.epilepsy.org.au**

Epilepsy Association Australia: A growing site with information offered separately for adults, children, and teens. Good general information for everyone and offers of many services for people in Australia.

32. **www.epilepsy.org.uk/**

the website of Epilepsy Action, the working name of the British Epilepsy Association.

33. **www.epilepsysupport.org.uk/**

Epilepsy Support: A British site offering patient-to-patient community support.

34. **www.epinet.org.au/**

The website of the Epilepsy Foundation of Victoria, Australia. You can participate in rating the websites included in the interesting list of links here, including some in other languages. Also some good practical tips.

35. **www.ibe-epilepsy.org/**

the website of the International Bureau for Epilepsy, a European-based group dedicated to improving the non-medical aspects of life with epilepsy around the world. The site also includes links to groups in many countries.

36. **www.ilae-epilepsy.org**

International League Against Epilepsy: A preeminent association of physicians and other health professionals working towards a world where no persons' life is limited by epilepsy. This website contains plenty of helpful information on epilepsy.

Research

1. **www.aesnet.org/**
   Website of the American Epilepsy Society, which promotes research and education for health care professionals.

2. **www.centerwatch.com/patient/studies/CAT62.html**
   Center Watch: A comprehensive list of epilepsy clinical studies going on throughout the United States.

3. **www.ninds.nih.gov/index.htm**
   the site of the National Institute of Neurological Disorders and Stroke, the part of the U.S. National Institutes of Health that covers brain disorders.

News about research developments and clinical trials.

Epilepsy support Groups

1. **http://www.chrissyandfriends.org/**
   Non-Profit organization that has support groups for children and teens with epilepsy and parents of children with epilepsy support groups in the NYC area.

2. **http://www.epilepsypenarth.ik.com**
   A support group situated in South Wales, open to anyone with epilepsy to join for help and support.

3. **http://dspace.dial.pipex.com/epilepsybereaved/**
   Information about SUDEP and other epilepsy deaths. Bereavement support and information on SUDEP research. Free booklet for young people with epilepsy.

4. **http://www.stixdesigns.com.au/epilepsy-support/index.html**
   Includes information, links and articles relating to epilepsy and related topics.

5. **http://www.epilepsysupport.org.uk/**
   Maintained by someone with the disease, this site offers support, advice, and useful links.

6. **http://groups.yahoo.com/group/ketogenic/**
   This is a support group for parents, family members, and caregivers who are using the Ketogenic Diet as a means of control for epilepsy or seizure disorders.

7. **http://groups.yahoo.com/group/childhoodepilepsy/**
   This mailing list is intended to support parents and other caregivers of children diagnosed with a form of epilepsy.

8. **http://groups.yahoo.com/group/parents-of-kids-with-epilepsy/**
   This Group is for Parents of Children With Epilepsy.

9. **http://home.ease.lsoft.com/Archives/Epilepsy-L.html**
   An E-mail based support group for people with epilepsy and those who care about and for them.

10. **http://groups.msn.com/laforabodydiseaseresearchtalk**
    This is a site about lafora body disease it has a message board and a links page.

11. **http://groups.yahoo.com/group/Epilepsy_Advocates/**
    Our group is made up of people that suffer from Epilepsy, parents of children with Epilepsy, people whom have previously had Epilepsy and friends of

those suffering from Epilepsy.

12. **http://groups.yahoo.com/group/epilepsy/**
For people with or know someone with Epilepsy. A place to talk, share stories, and for moral support. All ages welcome.

13. **http://groups.yahoo.com/group/epilepsy-kids/**
Epilepsy-Kids offers support and information for parents of children diagnosed with any form of epilepsy.

14. **http://groups.yahoo.com/group/KuekidsAustralia/**
The Kuekids Australia mailing list is to help parents, families, and friends of children using the ketogenic diet who may also have food intolerance problems.

15. **http://www.healthboards.com/epilepsy/**
Healthboards.com message board for information and support on epilepsy.

16. **http://www.orgsites.com/mn/epilepsy**
A guided independent living experience designed to prepare adults diagnosed to live and work on their own. Includes services, history, gallery, opportunities, and FAQs.

17. **http://www.users.bigpond.com/epilepsysupportgroup/Default.htm**
Offers information, events, issues, links, and contact details.

18. **http://www.cymepilepsysbytygwyn.co.uk/**
    Epilepsy Support Group in Wales, UK.

19. **http://au.groups.yahoo.com/group/australian_epilepsy_group/**
    Yahoo! Group for giving support and sharing information.

20. **http://www.debraboard.org/cgi-bin/temp.pl**
    A forum for people affected by this disorder.

21. **http://s4.invisionfree.com/VNS_Message_Board/index.php**
    A message board designed for information exchanged related to the Vagus Nerve Stimulator.

22. **http://groups.yahoo.com/group/Christian-Epilepsy-Support/**
    Christian oriented support for people with Epilepsy or family and friends of people with Epilepsy.

23. **http://groups.yahoo.com/group/Dating-and-Married-Life-with-Epilepsy/**
    Made for people to discuss issues that come up when they are dating and married, including handling different situations, embarrassment, children, and fear.

24. **http://www.pokwe.org.uk**
    A support web site for parent is of children with Epilepsy. Conversations via our mailing lists and

web content. This group was designed by parents for parents.

25. **http://groups.yahoo.com/group/epilepsycircle/**
This support group is for all who have an interest in epilepsy, whether it is for you, a family member, a friend, pupil, or a child that has a seizure disorder or epilepsy.

26. **http://groups.yahoo.com/group/adultswithepilep sy/**
This club is for adults (those 18 and older) to discuss their experiences with epilepsy, including causes, treatment and coping skills.

27. **http://groups.yahoo.com/clubs/epilepsysupportg roup**
Supportive club to help those with this disease. Includes free membership, message boards, and chat room. Also joining instructions.

28. **http://www.esgs.org.uk**
Offers an extensive FAQ's and a section talking about the different drugs to control epilepsy.

29. **http://www.wsm.inuk.com/epilepsy.htm**
Provides information about coping with this disease, seizures and convulsions. Details about the various types, and provides free help and advice using its email and telephone help lines.

30. **http://www.geocities.com/epilepsy911**
Providing worldwide support for people with

epilepsy. Includes FAQ's, links, support groups, and a pen pal site.

31. **http://groups.yahoo.com/group/ChristianEpilepsyPrayerGroup/**
    Are you a Christian living with or caring for someone with epilepsy? ...

32. **http://groups.yahoo.com/group/seizurealertdogs/**
    This group unites people who benefit from seizure alert or assistance dogs as well as people with epilepsy who benefit from a dog's companionship. We also Welcome people who want to learn more about these wonderful dogs.

33. **http://groups.msn.com/doosesyndrome**
    This is a group about doose and it has links to other websites but almost all information can be seen by non members of the group.

34. **http://groups.yahoo.com/group/EpilepsySurgeryGroup/**
    We deal with those who have had brain surgery, Vagus Nerve Stimulator surgery, and Gamma Knife Radio surgery. Our members are here to help answer any questions you may have. All are welcome, whether you are a surgery candidate or not.

35. **http://groups.yahoo.com/group/real_epilepsy/**
    A group based in Indiana. Epilepsy and seizure

disorder resources and support.

36. **http://groups.yahoo.com/group/RareEpilepsySyndromes/**
Rare epilepsy is very difficult on all families involved. This group has been made to help educate and support these families. Rare epilepsy's include. ESES/CSWS, LKS, LGS, Begnign Rolandic, and any other hard to recognize epilepsy.

37. **http://groups.yahoo.com/group/epilepsychat/**
A Place For Chat, Info, Support And To Have Fun

38. **http://www.pokweuk.org**
A UK based supporting website containing epilepsy information, children's stories, and a forum for parents of children with this disorder.

39. **http://groups.yahoo.com/group/TLE/**
Discussion of temporal lobe epilepsy and/or partial complex seizure disorder. For sufferers and friends/family of sufferers.

40. **http://beehive.thisisessex.co.uk/epilepsy**
Local branches of national organization providing support, advice and information on living with epilepsy

41. **http://health.groups.yahoo.com/group/doosesyndrome/**
For parents of children who have Doose Syndrome rare, severe, intractable epilepsy of early childhood. Group provides support for parents

along with insight into management of this disorder.

42. **http://groups.yahoo.com/clubs/epilepsy_surgery**
Details about this group which is for those who have thought of (or had) having brain surgery or Vagus Nerve Stimulator surgery. Registration required.

43. **http://groups.yahoo.com/group/worldofepilepsy/**
For people around the globe who have epilepsy and would like to share their recent or previous experience with epilepsy.

44. **http://groups.yahoo.com/group/wivesofmenwith epilepsy/**
It is the hope of this club to address problems, offer support, and encouragement to the wife whose husband has epilepsy.

45. **http://community.healingwell.com/community/?f=23**
Message boards and chat dedicated to sharing information, offering support, and coping strategies to manage this disorder. Community moderated by volunteers.

46. **http://groups.yahoo.com/group/epilepsycured/**
We are people sharing information on curing / preventing epilepsy - seizures. This is a support group for patients, parents, children, family members ... anyone and everyone.

47. **http://groups.yahoo.com/group/doosesyndrome/**
For parents of children who have, or who may have Myoclonic-Astatic Epilepsy, otherwise known as Doose Syndrome, a rare, severe, intractable epilepsy of early childhood.

48. http://dir.groups.yahoo.com/dir/Health___Wellness/Support/Diseases_and_Conditions/Epilepsy?

49. **http://groups.yahoo.com/group/epilepsysupportandeducation/**
We are a growing community to help bring people with epilepsy from all corners of the world to share experiences, thoughts, and provide information. All ages are welcome and you do not have to be epileptic to join.

50. **http://groups.yahoo.com/group/lennox-gsibs/**
Lennox Gastaut is a severe form of Epilepsy. This group is for the siblings (brothers & sisters) of those affected by LGS, or a seizure disorder.

51. **http://groups.yahoo.com/group/epilepsysupport4kids/**
This group is a support group for kids with epilepsy, kids interested in improving their condition by natural means.

52. **http://groups.yahoo.com/group/ToddlerswEpilepsy/**
This is a group for parents with toddlers that have Epilepsy. Parents with infants are welcome too.

53. **http://neuro-mancer.mgh.harvard.edu/cgi-bin/forumdisplay.cgi?action=topics&forum=Epilepsy&number=33**

54. **http://www.inspirationallivingonline.com**

A Mothers Voice

Written By Jenna Martin, ***Senior Editor***

Living life with epilepsy can be a colossal struggle. However, if you are Stacey Chillemi, it is a challenge and an opportunity to help others.

Stacey Chillemi is a mother of three, a wife, and writer. Her journey and reason for being is defined each day by the happiness in her children's eyes and the people with epilepsy she has helped through her writing. "Through this experience with epilepsy I have learned to accept my limitations and to change the way I look at things. Through my writing I am able to help others and just knowing I've helped is enough of a reward," said Chillemi.

How it Began

At five years of age, Chillemi contracted encephalitis from what doctors' surmise began as an ear infection. For four days, she lay in a coma and doctors were unsure of whether she would suffer from paralysis as well as the extent of the brain damage. Fortunately, she recovered from her bout with encephalitis with no paralysis. However, she was left with epilepsy. Since her diagnosis, 27 years ago, Chillemi has had seizures ranging in severity from mild seizures in her sleep to tonic-clonic seizures.

Living with Limitations

As a mother of three and a woman with epilepsy, Chillemi is realistic about her limitations, "Having epilepsy and being a mom is difficult at times because I worry that if I have a seizure, and I am unable to recover fast enough, that my kids will suffer," said Chillemi. In fact, the entire time Chillemi has been a mother she has had only one tonic-clonic seizure resulting in serious injury. According to Chillemi, she was walking the dog with her children when she felt a seizure coming on. She immediately instructed the children to go inside and wait downstairs for her. Following the seizure, Chillemi realized she had suffered a head injury and reached out to a neighbor for help. After the tonic-clonic seizure, Chillemi decided to write a children's book, called "My Mommy Has Epilepsy." Her goal was to help children understand epilepsy in an age appropriate very as well as to help dispel some of the fear she had witnessed her own children experience. "I don't want my children to get nervous or to worry about my seizures and the tonic- clonic seizure really motivated me to write a children's book to help them and other kids cope and understand epilepsy."

She admits she is also limited by not being able to drive, but attributes her ability to ask for help when she needs it as one more lesson learned. "At first it was difficult to rely on other people to drive the children and me places. I felt bad asking family and friends. But, now I've accepted my limitations and accepted who I am."

Wisdom for Women

Chillemi cautions women with epilepsy to monitor their stress level. "Don't try to accomplish too much. Do as much as you can and remember to set realistic goals and to reward yourself each day." She also believes that in order to live with epilepsy and maintain a positive attitude it is important to focus on yourself. " Don't look at what other people can do, only focus on yourself and your abilities because if you constantly compare, then you are adding to your stress level, which leads to being more physically drained and ultimately leaves you open to experience more seizures." She advises women with epilepsy to educate themselves about their epilepsy and believes knowledge helps alleviate stress as well, "Women need to consult with their doctor, but not rely on their doctor for all of their information. In order to really feel empowered it is essential to take the initiative to learn all you can about epilepsy."

Meet the Author

"Author Stacey Chillemi"

I. always loved to create journals when I was young, but if you asked me then, "What I wanted to be when I grow up" I would not have said a writer.

When I was five, I developed epilepsy. In college, I was confronted with many different obstacles concerning my disorder. Confused, not knowing where my path was leading me and where my journey was taking me I decided to go to bookstores and libraries to search for answers about the questions I had about my disorder, "Epilepsy."

At that time, there were few books on the bookshelf about epilepsy. Doctors wrote all the books. The books were written in medical terminology, so if you were not educated in the medical field, you had no idea what they were trying to explain.

I was angered by this, so I decided to do my own research about epilepsy and write my own book to help others with the disorder so they could get the answers to the questions they had.

I wrote an article asking people with epilepsy to write to me and tell me their story. Hundreds of letters swarmed in. I interviewed about 400 people. I learned so much from reading other people's stories and talking to others.

Writing made me feel so good. I helped myself, I helped others, and the satisfaction of helping others was tremendous. I wrote my first book, "Epilepsy You're Not

Alone" in 2000 and since then I have not stopped writing.

EPILEPSY

The Most Important Secrets You Must Learn In Order To Live, Learn, and Be Happy With Epilepsy

Author's Background

"Author Stacey Chillemi"

New Jersey-based freelance writer, providing worldwide

Summary: Stacey Chillemi graduated from Richard Stockton College in Pomona, New Jersey, majoring in marketing and advertisement. In the mid-nineties while in college, she began her first book, Epilepsy: You're Not Alone. It was published six years later. Before and after graduation in 1996, she worked in New York City for NBC. Since the birth of her children, she has been a freelance journalist.

She has written features for journals and newspapers. Her articles have appeared in dozens of newspapers and magazines in North America and abroad. She won an award from the Epilepsy Foundation of America in 2002 for her help and dedication to people with epilepsy.

Accomplishments:

- Writer for Neurology Now Magazine (The Academic Academy of Neurology – The Epilepsy Column) February 2010

- February 2010, Wrote an article about Epilepsy & Menstruation with Dr. Devinsky (Epileptologist from NYU)
- Book Signing at Borders in Freehold, New Jersey for Faith, Courage, Wisdom, Strength and Hope" - July 2009
- H.O.P.E. Mentor, for the Epilepsy Foundation
- Speaker at different events for schools, organizations, political events
- Spoke in front of Congress in Washington for employment discrimination for people with Epilepsy
- Appeared on four talk shows to discuss epilepsy focusing on the importance of understanding epilepsy, how to help someone having a seizure and giving people with epilepsy encouragement and hope for the future.
- Appeared on radio stations discussing epilepsy
- Appeared on the Michael Dressor Show - Health Radio
- Appeared in newspapers all over New Jersey such as, The Leader,

Belleville Post and the Star Ledger.

- Received awards in my achievements and certificates in recognition for outstanding efforts in trying to improve society.
- Active participant in organizations and activities.
- Published over 400 articles
- Author has a dynamic personality and strong public speaking skills.

BOOKS PUBLISHED BY STACEY CHILLEMI:

- The Complete Herbal Guide: A Natural Approach to Healing the Body
- How to Live Comfortably with Asthma
- Epilepsy You're Not Alone
- Eternal Love: Romantic Poetry Straight from the Heart
- My Mommy Has Epilepsy (Children's Book)
- My Daddy Has Epilepsy (Children's Book)
- Keep the Faith: To Live and Be Heard from the Heavens Above (poetry book)

- Live, Learn, and Be Happy with Epilepsy
- Epilepsy and Pregnancy: What Every Woman Should Know Co-authored by Dr. Blanca Vasques
- Faith, Courage, Wisdom, Strength and Hope
- How to Be Wealthy Selling Informational Products on the Internet
- How to Become Wealthy in Real Estate
- How to Become Wealthy Selling Ebooks
- Life's Missing Instruction Manual: Beyond Words
- How To Become Wealthy Selling Products on The Internet
- Breast Cancer: Questions, Answers & Self-Help Techniques
- How Thinking Positive Can Make You Successful: Master The Power Of Positive Thinking

STACEY CHILLEMI STORIES AND POETRY HAVE BEEN PUBLISHED IN:

- Chicken Soup for the Recovering Soul
- Chicken Soup for the Shoppers Soul
- Whispers of Inspiration

NBC
Network
New York City

- Worked for NBC on Dateline
- Worked for Channel 4 News
- Worked for the Today Show

The Journal Magazine
Journalist

Wrote comments on topics of reader interest to stimulate and mold public opinion, in accordance with viewpoints and policies of publication.

- Prepared articles from knowledge of topic and editorial position of publication, supplemented by additional study and research.
- Recommended topics and position to be taken by publication on specific public issues.

UZURI Fashion
Magazine

Managing

Editor

- Planned, coordinated and directed editorial activities and formulated policy.
- Supervised workers who assisted in the selecting and preparing of material for publication.
- Formulated policy, coordinated department activities, established production schedules and solved publication problems.
- Wrote and assigned staff members and freelance writers to write articles, reports, editorials, reviews and other material.

Awards: June 26, 2002, I was honored an award by the Epilepsy Foundation of New Jersey for Outstanding Volunteer Award.

Web sites:

http://www.authorsden.com/staceydchillemi

http://www.stores.lulu.com/staceychil

http://freelanceinternational.viviti.com

EPILEPSY

The Most Important Secrets
You Must Learn
In Order To Live, Learn, and Be Happy With Epilepsy

References

Adapted from: Schachter SC, editor. *Brainstorms: epilepsy in our words.* New York: Raven Press; 1993; and Schachter SC, editor. *The brainstorms companion: epilepsy in our view.* New York: Raven Press; 1995.

Steven C. Schachter, M.D.

Schachter SC (2003). Epilepsy: Etiology and manifestations. In RW Evans, ed., Saunders Manual of Neurologic Practice, part VII, pp. 244–265. Philadelphia: The Curtis Center.

Blume WT (2003). Diagnosis and management of epilepsy. Canadian Medical Association Journal, 168(4): 441–448.

Pedley TA, et al. (2000). Epilepsy. In LP Rowland, ed., Merritt's Neurology, 10th ed., pp. 813–833. Philadelphia: Lippincott Williams and Wilkins.

Brodie MJ, French JA (2000). Management of epilepsy in adolescents and adults. Lancet, 356: 323–329.

Kwan P, Brodie MJ (2000). Early identification of refractory epilepsy. New England Journal of Medicine, 342(5): 314–319.

Wiebe S, et al. (2001). A randomized, controlled trial of surgery for temporal-lobe epilepsy. New England Journal of Medicine, 345(5): 311–318.

American Academy of Neurology (2004). New treatment guidelines: Efficacy and tolerability of the new antiepileptic drugs, I: Treatment of new onset epilepsy. Available

online:
http://www.neurology.org/cgi/reprint/62/8/1252.pdf.

American Academy of Neurology (2004). New treatment guidelines: Efficacy and tolerability of the new antiepileptic drugs, II: Treatment of refractory epilepsy. Available online:
http://www.neurology.org/cgi/reprint/62/8/1261.pdf.

Diaz-Arrastia R, et al. (2002). Evolving treatment strategies for epilepsy. JAMA, 287(22): 2917–2920.

Hanhan UA, et al. (2001). Status epilepticus. Pediatric Clinics of North America, 48(3): 683–694.

Marson A, Ramaratnam S (2002). Epilepsy. Clinical Evidence (8): 1313–1328.

See H. Reisner, ed. Children with Epilepsy (1988); R. J. Gunnit, Living Well with Epilepsy (1990); O. Devinsky, A Guide to Understanding and Living with Epilepsy (1994); publications of the Epilepsy Foundation of America.

Schachter SC (2003). Epilepsy: Etiology and manifestations. In RW Evans, ed., Saunders Manual of Neurologic Practice, part VII, pp. 244–265. Philadelphia: The Curtis Center.

Blume WT (2003). Diagnosis and management of epilepsy. Canadian Medical Association Journal, 168(4): 441–448.

Pedley TA, et al. (2000). Epilepsy. In LP Rowland, ed., Merritt's Neurology, 10th ed., pp. 813–833. Philadelphia: Lippincott Williams and Wilkins.

Brodie MJ, French JA (2000). Management of epilepsy in adolescents and adults. Lancet, 356: 323–329.

Kwan P, Brodie MJ (2000). Early identification of refractory epilepsy. New England Journal of Medicine, 342(5): 314–319.

Wiebe S, et al. (2001). A randomized, controlled trial of surgery for temporal-lobe epilepsy. New England Journal of Medicine, 345(5): 311–318.

American Academy of Neurology (2004). New treatment guidelines: Efficacy and tolerability of the new antiepileptic drugs, I: Treatment of new onset epilepsy. Available online: http://www.neurology.org/cgi/reprint/62/8/1252.pdf.

American Academy of Neurology (2004). New treatment guidelines: Efficacy and tolerability of the new antiepileptic drugs, II: Treatment of refractory epilepsy. Available online: http://www.neurology.org/cgi/reprint/62/8/1261.pdf.

Diaz-Arrastia R, et al. (2002). Evolving treatment strategies for epilepsy. JAMA, 287(22): 2917–2920.

Hanhan UA, et al. (2001). Status epilepticus. Pediatric Clinics of North America, 48(3): 683–694.

Marson A, Ramaratnam S (2002). Epilepsy. Clinical Evidence (8): 1313–1328.

Jarrar RG, Buchhalter JR (2003). Therapeutics in pediatric epilepsy, part 1: The new antiepileptic drugs and the ketogenic diet.

Mayo Clinical Procedures, 78: 359–370.

Schachter SC (2002). Vagus nerve stimulation therapy summary: Five years after FDA approval. Neurology, 59(4): S15–S20.

Buchhalter JR, Jarrar RG (2003). Therapeutics in pediatric epilepsy, part 2: Epilepsy surgery and vagus nerve stimulation. Mayo Clinic Proceedings, 78: 371–378.

Helmers SL, et al. (2001). Vagus nerve stimulation therapy in pediatric patients with refractory epilepsy: Retrospective study. Journal of Child Neurology, 16(11): 843–848.

[1] Data on file, Novartis Pharmaceuticals Corporation.
[2] Guerreiro MM, Vigonius U, Pohlmann H, et al. A double blind controlled clinical trial of Oxcarbazepine versus Phenytoin in children and adolescents with epilepsy. Epilepsy Res. 1997; 27:205-213.
[3] Bill PA, Vigonius U, Pohlmann H, et al. A double blind controlled clinical trial of Oxcarbazepine versus Phenytoin

in adults with previously untreated epilepsy. Epilepsy Res. 1997; 27:195-204.

[4] Beydoun A, Sachdeo RC, Rosendfeld WE, et al. Oxcarbazepinemonotherapy for partial-onset seizures; a multicenter, double blind, clinical trial. Neurology. 2000; 54:2245-2251.

Brodie M, Kwan P (2002). Staged approach to epilepsy management. Neurology, 58(8, Suppl 5): S2–S8.

Jarrar RG, Buchhalter JR (2003). Therapeutics in pediatric epilepsy, part 1: The new antiepileptic drugs and the ketogenic diet. Mayo Clinical Procedures, 78: 359–370.

The Cleveland Clinic
Department of Patient Education and Health Information
9500 Euclid Ave. NA31 Cleveland, OH 44195
216/444-3771 or 800/223-2273 ext.43771
healthl@ccf.org

ISBN: 978-1-300-55340-3

CPSIA information can be obtained
at www.ICGtesting.com
Printed in the USA
BVOW06s0825201216
471338BV00001B/96/P

9 781300 55340